ACS SYMPOSIUM SERIES **749**

Olefin Polymerization

Emerging Frontiers

Palanisamy Arjunan, EDITOR
Exxon Chemical Company

James E. McGrath, EDITOR
Virginia Polytechnic Institute and State University

Thomas L. Hanlon, EDITOR
Albemarle Corporation

American Chemical Society, Washington, DC

Library of Congress Cataloging-in-Publication Data

Olefin polymerization : emerging frontiers / Palanisamy Arjunan, James E. McGrath, Thomas L. Hanlon, editors.

    p.   cm.—(ACS symposium series ; 749)

    Includes bibliographical references and index.

    ISBN 0–8412–3614–3

    1. Polyolefins—Congresses. 2. Polymerization—Congresses.

    I. Arjunan, Palanisamy. II. McGrath, James E. III. Hanlon, Thomas L. IV. Series.

TP1180.P67 O44 1999
668.4′234—dc21   99–47688

The paper used in this publication meets the minimum requirements of American National Standard for Information Sciences—Permanence of Paper for Printed Library Materials, ANSI Z39.48–1984.

PRINTED IN THE UNITED STATES OF AMERICA

Foreword

THE ACS SYMPOSIUM SERIES was first published in 1974 to provide a mechanism for publishing symposia quickly in book form. The purpose of the series is to publish timely, comprehensive books developed from ACS sponsored symposia based on current scientific research. Occasionally, books are developed from symposia sponsored by other organizations when the topic is of keen interest to the chemistry audience.

Before agreeing to publish a book, the proposed table of contents is reviewed for appropriate and comprehensive coverage and for interest to the audience. Some papers may be excluded in order to better focus the book; others may be added to provide comprehensiveness. When appropriate, overview or introductory chapters are added. Drafts of chapters are peer-reviewed prior to final acceptance or rejection, and manuscripts are prepared in camera-ready format.

As a rule, only original research papers and original review papers are included in the volumes. Verbatim reproductions of previously published papers are not accepted.

ACS BOOKS DEPARTMENT

Contents

PROGRESS IN CHARACTERIZATION

INDEXES

Preface

Despite the fact that polyolefins were one of the last types of polymers to be commercialized, they have become the most widely used polymers worldwide. Polyolefins, including polyethylene, polypropylene, various olefin copolymers, and polydienes comprise much more than half of the volume of polymeric materials produced in the United States each year. Major advances in this field have continued during the past decade, in contrast to many other areas of industrial research development. Undoubtedly, this advancement is related to the economic importance and availability of high-volume petrochemical monomers, such as ethylene, propylene, and other higher α-olefins, as well as emerging discoveries in olefin polymerization processes and catalysis. Many research activities, such as developing new catalysts, understanding polymerization mechanisms, modifying the products, and improving the physical properties of the materials have been proceeding at a very fast pace, especially in the industrial laboratories around the world.

It is important and exciting to bring researchers who are from both the academic and industrial sectors, who are active in this area together, to present their new findings. To the best of our knowledge, a symposium covering the diverse aspects of olefin polymerization has not been held in the United States for many years. With this in mind, a symposium entitled "Advances in Olefin Polymerization" was organized at the 215th American Chemical Society (ACS) National Meeting in Dallas, Texas from March 29 to April 2, 1998. This symposium covered both scientific and technological aspects of olefin polymerization that were included in four sessions: progress in catalysis, progress in polymerization, progress in polymer design, and progress in mechanism–characterization. More than twenty papers, including six foreign contributions, were presented during the two and a half days of the symposium. It was truly exciting to see renowned researchers from both academic and industrial laboratories present their key discoveries and share the important, emerging developments in olefin polymerization.

This book is intended to capture some of the most recent and emerging technical achievements presented in the above symposium. It is certainly impossible to cover every aspect of this fast-paced research area. Instead, the twelve chapters in this volume provide a balanced coverage of the key developments in four different emerging frontiers of olefin polymerization: progress in catalysis, progress in polymerization, progress in polymer design, and progress in characterization.

The first section includes key discoveries in catalysis, specifically doubly-silylene bridged group 3 and group 4, C_5, and C_1, symmetric metallocenes for polypropylene and polypentene, lanthanides for polybutadienes, and siloxy functionalized group 4 metallocenes for polyethylene, polypropylene, and copolymers thereof. The polymerization mechanisms, the microstructure of products, and the role of catalyst structure in controlling the above features are also described.

In the second section, recent progress in polymerization, with special reference to propylene polymerization using supported magnesium chloride catalysts, homo-and copolymerization of styrene by bridged benzindenyl zirconocene complex, and random copolymerization of propylene and styrene with monocyclo-penta-dienyl titanium–MAO catalyst are discussed. The effect of catalyst–cocatalyst, polymerization conditions on the stereo- and regiospecificity of the products and reactivity ratios are also well explained.

The third section contains current efforts in designing new polyolefin architectures, specifically metallocene-based polyethylenes, functionalized polyethylenes and copolymers thereof. Improved property–performance attributes of the above materials by controlling their molecular–structural characteristics such as molecular weight distribution, comonomer type–content–distribution, and branching level are described. A novel approach to functionalized polyolefin structures is exemplified by incorporating the *p*-methylstyrene comonomer in various polyolefin backbone structures.

Generally, the key discoveries in olefin polymerization would not have happened without the aid of new characterization techniques and tools that are currently available. Deservingly, the final section is devoted to various new characterization techniques such as variable temperature solid-state ^{13}C NMR, ^{1}H NMR, dynamic mechanical analysis, and temperature-rising elution fractionation techniques. The structural distributions in Ziegler–Natta catalyst-based EP copolymers are elucidated by applying the Doi two-state statistical fit of their pentad–heptad, *meso*, and *racemic* sequence distributions. The use of variable temperature, solid-state ^{13}NMR methods to determine the crystal structure, interfaces, and branching of various linear low-density polyethylenes is also included. A novel approach to determine the short-chain branching distribution for amorphous ethylene copolymers is illustrated by applying both dynamic mechanical analysis and temperature-rising elution fractionation technique. New approaches for characterization of methylaluminoxanes and determination of trimethylaluminum content by using the ^{1}H NMR technique are also described.

As we know, the history of this field goes back many years. Previous reviews (*1–4*) provide sources for the early developments. Some recent developments, which are further amplified in this volume, are also reported by others (*5–10*). However, the integrated effort of this book contrasts with many other reviews that have a narrower focus. The symposium proceedings are written as a full manuscript version of the presentations, and the volume should be of great interest to both industrial and academic researchers in the area of olefin polymerization. With the organization of the symposium and the publication of this book, we hope to present the key discoveries in the emerging frontiers of olefin polymerization to a general audience.

Acknowledgments

It was a pleasure and an honor to organize the symposium and edit this book. We thank the authors for their fine contributions. We thank the Program Committee of the ACS Division of Polymer Chemistry, Inc. for providing us the opportunity to organize the symposium. We also gratefully acknowledge the generous financial support of the following cosponsors of the symposium: ACS Division of Polymer Chemistry, Inc., ACS Corporate Associates, ACS Petroleum Research Fund, Amoco Chemical Company, Baxter International Corporation, BF Goodrich Company, Dow-Corning Corporation, DuPont, Exxon Chemical Company, PQ Corporation, and Univation Technologies. The financial support was used to enable both the foreign and U.S. speakers to participate in the symposium.

Finally, we thank both Anne Wilson and Kelly Dennis, ACS Books Department, who kept us on task in a firm but patient manner. Their interest and assistance in publishing this book are greatly appreciated.

References

1. Clark, A.; Hogan, J. P.; Banks, R.; Lanning, W. *Ind. Eng Chem.* **1956,** *48,* 1152.
2. Vandenberg, E. J.; Repka, E. *Polymer Processes;* Schildknecht, C. E., Ed.; Wiley: New York; 1977.
3. Boor, J. *Ziegler–Natta Catalysts and Polymerization;* Academic: Orlando, FL; 1979.
4. Soga, K.; Shiono, T. *Prog. Polym. Sci.* **1997,** *22,* 1503–1546.
5. Kaminsky, W. K.; Arndt, M. "Metallocenes for Polymer Catalysis", *Adv. Polym. Sci.* **1997,** *127,* 143–187.
6. Locatelli, P. "Ziegler–Natta Catalysts: No End in Sight to Innovation", *Trends Polym. Sci.* **1966,** *4(10),* 326.
7. Hamiele, A. E.; Soares, J. B. P. "Polymerization Reaction Engineering Meatallocene Catalysts", *Prog. Polym. Sci.* **1996,** *27,* 651–676.
8. Chung, T. C. *New Advances in Polyolefins;* Plenum: New York; 1993.

9. Benedikt, G. M.; Goodall, B. L. *Metallocene Catalyzed Polymers;* Plastics Design Library; 1998.
10. Brotovsek, G. J. P.; Gibson, V. C.; Wass, D. F. *Angew. Chem. Int. Ed.* **1999,** *38,* 428–447.

PALANISAMY ARJUNAN
Exxon Chemical Company
5200 Bayway Drive
Baytown, TX 77520–2101

JAMES E. MCGRATH
Department of Chemistry
Virginia Polytechnic Institute and State University
Blacksburg, VA 24061–0344

THOMAS L. HANLON
Albemarle Corporation
8000 GSRI Avenue
Baton Rouge, LA 70820

PROGRESS IN CATALYSIS

Chapter 1

Reactivity and Mechanistic Studies of Stereocontrol for Ziegler–Natta Polymerization Utilizing Doubly-Silylene Bridged Group 3 and Group 4 Metalocenes

D. L. Zubris, D. Veghini, T. A. Herzog, and J. E. Bercaw[1]

Arnold and Mabel Beckman Laboratories of Chemical Synthesis, California Institute of Technology, Pasadena, CA 91125

A family of zirconocene, yttrocene, and scandocene complexes have been prepared utilizing an easily modified doubly bridged ligand array, $(1,2-SiMe_2)_2\{\eta^5-C_5H_2-4-R\}\{\eta^5-C_5H-3,5-(CHMe_2)_2\}$, developed in our laboratories. Variations in ligand array substitution and metal center provide complexes that exhibit a range of stereoselectivities and activities in α-olefin polymerizations. A series of C_s- and C_1- symmetric metallocenes have been utilized for the preparation of polypropylene and polypentene under various polymerization conditions. Analysis of the resulting polymer microstructures has provided valuable information about the factors that dictate syndiospecificity, as well as common error forming mechanisms for these systems.

The use of metallocenes as catalysts for the polymerization of ethylene and α-olefins has received increasing attention due to the striking activities and stereoselectivities they can effect (*1*). These catalyst systems are either neutral species of the form Cp'$_2$M-R (M = group 3 transition metal or lanthanide; Cp' = $(\eta^5-C_5H_5)$ or substituted cyclopentadienyls) or cationic species of the form [Cp'$_2$M-R]$^+$X$^-$ (M = group 4 transition metal; X$^-$ = non-coordinating anion). Efforts have been directed towards elucidating the active species and mechanism of polymerization to help determine the source of the remarkable regio- and enantioselectivities found with these systems. Slight variations in metallocene structure provide access to a variety of polymer microstructures, molecular weights, and melting temperatures; some of these microstructures are illustrated in Figure 1.

Catalyst symmetry and the microstructure of the resulting poly-α-olefin are related in a rational manner. C_2- symmetric metallocenes, such as Brintzinger's chiral *ansa*-zirconocene dichlorides (*2*), can be activated with methylaluminoxane (MAO) to produce highly isotactic polypropylene (PP), as depicted in Figure 2.

[1]Corresponding author.

Isotactic polymer arises from the metallocene catalyst selectively enchaining one of the olefin enantiofaces. A family of catalysts that exhibit either C_s- or C_I-symmetry have been developed by Ewen and Razavi (*3, 4*) using a substituted, linked cyclopentadienyl/fluorenyl ligand array. Unlike C_2- symmetric systems, C_s-symmetric catalysts enchain α-olefins with regularly alternating enantiofaces and produce syndiotactic polymer. Minor variations in ligand substitution, as illustrated in Figure 3, provide catalysts that are capable of producing polypropylenes from moderately syndiotactic, through "hemi-isotactic" to moderately isotactic.

Characteristics of Syndiospecific Metallocene Systems

The relationship between the substitution and symmetry of the cyclopentadienyl ligand framework, the pendant polymer chain, and the α-olefin alkyl has been examined for both iso- and syndio- specific systems. Nevertheless, the important stereodirecting interactions in the transition state that induce a highly stereospecific polymerization are not fully understood. In recent years, we have probed the criteria necessary for generating a syndiospecific metallocene catalyst system. Syndiospecificity is believed to arise from three key features: (1) the C_s- symmetric *ansa*-metallocene has one cyclopentadienyl possessing bulky substituents flanking the center of the metallocene wedge to direct the polymer chain segment (up) toward the less substituted cyclopentadienyl, whether it resides on the left or right side in the transition structure for propagation (Figure 4); (2) an open region between the bulky substituents on the lower cyclopentadienyl to accommodate the α-olefin alkyl (methyl for propylene), which is directed trans to the polymer segment (down); (3) migratory insertions that result in regular alternation of the monomer approach from the left and right sides of the metallocene wedge. The dominant steric interaction in the transition structure leads to the trans relationship between the polymer segment and the α-olefin alkyl (*5-8*).

With these features in mind, we developed a ligand array that could be easily modified to further probe the nature of syndiospecificity. Our laboratory recently reported the preparation of a series of C_s- and C_I- symmetric doubly bridged metallocenes and some preliminary polymerization data (*9*). A variety of zirconocenes can be prepared with this ligand array (Figure 5), and when activated with MAO or borate co-catalysts they serve as highly active, stereospecific catalysts for the polymerization of propylene and 1-pentene (Henling, L. M.; Herzog, T. A.; Bercaw, J. E. *Acta Cryst.*, **1996**, *C52*, in press) (Veghini, D.; Day, M.; Bercaw, J. E. *Inorganica Chimica Acta*, accepted).

The C_s- symmetric precatalysts, **1a-c**, display high activity and syndiospecificity when activated with methylaluminoxane. The C_I- symmetric precatalysts, **1d-e**, display markedly different polymerization behavior, switching from moderately syndiospecific to isospecific when propylene concentration is decreased. These observations demonstrate that seemingly minor changes in the ligand array (here, changing the R substituent) can have major implications for the stereoselectivity of a given metallocene. In addition this ligand framework can be incorporated into analogous yttrium and scandium compounds. In contrast to the

4

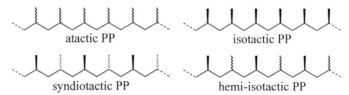

atactic PP   isotactic PP

syndiotactic PP   hemi-isotactic PP

Figure 1. Molecular structures of polypropylenes.

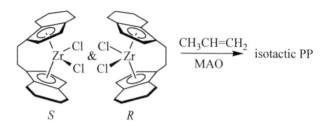

Figure 2. C_2- symmetric zirconocene catalysts produce isotactic polypropylene.

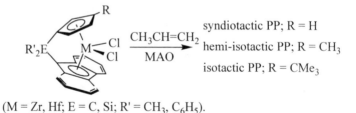

syndiotactic PP; R = H
hemi-isotactic PP; R = CH$_3$
isotactic PP; R = CMe$_3$

(M = Zr, Hf; E = C, Si; R' = CH$_3$, C$_6$H$_5$).

Figure 3. Metallocenes for the production of a range of polypropylene microstructures.

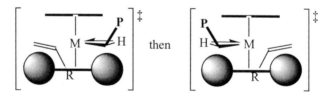

Figure 4. Transition structures for syndiospecific polymerization of α-olefins.

zirconocene systems, these group 3 metallocenes do not require activation to serve as polymerization catalysts. Synthesis of this family of compounds and subsequent polymerization studies provide a more complete understanding of the important factors for achieving syndiospecificity.

Catalyst Activity and Polymer Molecular Weights

Zirconocenes **1a-e**, in combination with MAO, exhibit high activities for polymerization of liquid propylene at 20 °C, as enumerated in Table I. These activities are higher and polymer polydispersities are generally lower than those of metallocenes such as $Ph_2C[(\eta^5-C_5H_4)(\eta^5-C_{13}H_8)]ZrCl_2$ (**2a**), that bear a linked cyclopentadienyl/fluorenyl ligand array.

Table I. Activity, tacticity, and molecular weight data for polypropylenes produced by 1a-e/MAO and 2a/MAO.

Catalyst	T(°C)	Activity [a]	[r] (%)	M_w	M_w/M_n
1a	20	2,800,000	>99.0	1,250,000	1.9
1a	50	7,900,000	96.6	330,000	2.3
1a	70	16,900,000	94.1	160,000	3.2
1b	20	320,000	99.4	980,000	2.0
1c	20	350,000	97.5	790,000	1.8
1d	20	250,000	75.4	190,000	2.2
1e	20	490,000	78.7	470,000	1.9
2a	20	130,000	99.2	840,000	2.3

[a] Defined as (g P / g Zr · hr)

Polymer molecular weights on the order of 1,000,000 with polydispersities of approximately 2 are found when the C_s- symmetric catalysts **1a-c** are employed in liquid propylene at 20 °C. As shown for catalyst **1a**, polymer molecular weights decrease and polydispersities increase with increasing reaction temperature. When propylene polymerizations are carried out in toluene solution, polymer molecular weight is found to increase with increasing propylene concentration, as depicted in Figure 6.

End group analysis, using ^1H NMR spectroscopy, (10) of low molecular weight polypropylene suggests that β-H elimination is the predominant chain termination pathway for this system. Further support is derived from a primary kinetic isotope effect k(β-H)/k(β-D) of 1.6 for chain termination, by comparison of the number average molecular weight of poly-d_0-propylene versus poly-2-d_1-propylene (Table II).

Table II: Polymer analyses for poly-d_0-propylene and poly-2-d_1-propylene produced by 1b/MAO.

Polymer	[r] (%)	[rmrr] (%)	[rmmr] (%)	M_n	M_w/M_n
poly-d_0-propylene	83.3	17.3	2.2	32,840	1.9
poly-2-d_1-propylene	84.2	19.6	1.6	56,830	1.8

Stereospecificity and Stereoerrors for C_s- Symmetric Systems

While superb syndiospecificity is observed when metallocenes 1a-c are used at high olefin concentrations (liquid propylene), errors in the polymer microstructure are increasingly present as olefin concentration is decreased. One error-forming mechanism of particular importance to syndiospecific systems is site epimerization, inversion of configuration at the metal center resulting from the polymer chain swinging from one side of the metallocene wedge to the other without insertion. Site epimerization disturbs the regularity of migratory insertions, and consequently provides errors in the polymer microstructure. For catalyst systems 1a-e, the balance between bimolecular chain propagation and monomolecular site epimerization provides variations in tacticity. Therefore, examining the effects of changes in temperature (as shown above for catalyst 1a) and olefin concentration can provide information about error forming mechanisms, such as site epimerization, for our C_s- and C_1- symmetric catalyst systems.

When propylene enchainment does not occur with perfectly alternating enantiofaces, two types of errors are observed: isolated m- and mm- stereoerrors. Three types of error forming mechanisms are considered for syndiospecific systems: (1) enantiofacial misinsertion, which provides a mm- triad (2) site epimerization, migration of the polymer chain without insertion, which provides an isolated m- diad and (3) chain epimerization, which can provide m- or mm- stereoerrors, dictated by the configuration at the metal center during the process. While the first process is independent of propylene concentration, the latter two exhibit a dependence on propylene concentration. The prevalence of these errors can be investigated by varying olefin concentration and examining polymer tacticity. An understanding of the source of errors also provides a more complete picture of the key stereocontrol mechanisms operating for these doubly bridged metallocenes.

Polymerization experiments were performed with 1a-c and MAO at various propylene pressures in toluene solution. The stereoselectivity, exemplified by [r], increases with increasing propylene concentration, suggesting that enantiofacial misinsertion is not a major stereoerror mechanism. The r- diad content at varying propylene concentrations is illustrated in Figure 7.

Examination of the stereoerror profile reveals that 2-3 % mm- triads are present at low propylene concentration for all three catalysts, and at increased olefin concentration are absent for 1b-c and account for less than 1% of the stereoerrors for 1a. Chain epimerization appears to be the major source of these mm- triads, with enantiofacial misinsertion accounting for less than 1% of these errors. Examination of isolated m- diads reveals that the amount of these

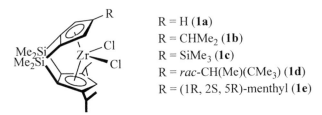

R = H (**1a**)
R = CHMe$_2$ (**1b**)
R = SiMe$_3$ (**1c**)
R = *rac*-CH(Me)(CMe$_3$) (**1d**)
R = (1R, 2S, 5R)-menthyl (**1e**)

Figure 5. Doubly-linked zirconocene catalysts for stereospecific polymerizations of α-olefins.

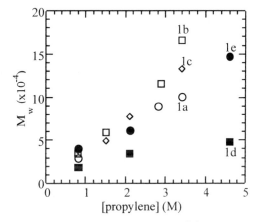

Figure 6. Dependence of molecular weights on propylene concentration for **1a-e**/MAO.

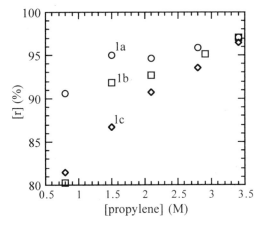

Figure 7. Dependence of *r*- diads on propylene concentration for C_s-symmetric catalysts **1a-c** /MAO.

stereoerrors decreases over the reported concentration range for catalysts **1a-c**. These changes in polymer microstructure with varying propylene concentrations are displayed in Figure 8.

To determine the contribution of chain epimerization to isolated *m*- diads, a labeling study analogous to those reported by Brintzinger (*11-12*) and Busico (*13-14*) was undertaken. Since the proposed mechanism for chain epimerization involves β-H elimination, one can elucidate the importance of this stereoerror mechanism by probing for a primary kinetic (or thermodynamic pre-equilibrium) isotope effect using appropriately labeled substrates. Parallel polymerizations of d_0-propylene and 2-d_1-propylene using **1b**/MAO as the catalyst system were carried out; microstructure analysis (by $^{13}C\{^1H\}$ NMR) shows essentially identical pentad distributions for both the labeled 2-d_1- (84.2 %) and the unlabeled d_0-propylene (83.3 %) substrates. Examination of the ^2D NMR reveals principally $-CD(CH_3)$ groups, with 2-3 % $-CH(CH_2D)$ groups that can be attributed to chain epimerization. This 2-3 % corresponds to the bulk of the *mm*- triads observed. Given that the amount of *m*- diads observed exceeds *mm*- triads at all conditions, it follows that site epimerization is the major error mechanism for these systems.

Group 3 C_s- Symmetric Systems

While C_s- symmetric zirconocenes utilizing the doubly bridged ligand array (1,2-$SiMe_2)_2\{\eta^5-C_5H_2-4-R\}\{\eta^5-C_5H-3,5-(CHMe_2)_2\}$ have further clarified the important characteristics of syndiospecific systems, the study of analogous yttrocene and scandocene complexes also can provide further mechanistic insight. Theoretical studies by Bierwagen et al. (*15*) predict that the energetic preference of an alkyl group to lie in the middle of the wedge for a C_s- symmetric group 3 metallocene will prevent these systems from forming syndiotactic polymer. This energetic preference would appear to be inconsistent with one of the necessary conditions for syndiospecific polymerization of α-olefins, which involves olefin approach from regularly alternating sides of the metallocene wedge. We therefore undertook the synthesis of a C_s- symmetric group 3 metallocene to ascertain whether or not a group 3, neutral metallocene with the (1,2-$SiMe_2)_2\{\eta^5-C_5H_2-4-R\}\{\eta^5-C_5H-3,5-(CHMe_2)_2\}$ ligand would yield syndiotactic polymer.

The *t*-butyl substituted doubly bridged ligand array has been used for the preparation of C_s- symmetric yttrocene and scandocene chloride complexes, as evidenced by NMR spectroscopy and X-ray crystallography. Both species

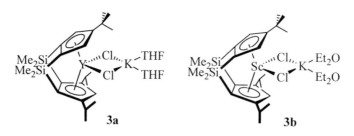

crystallize as potassium chloride adducts, with variable amounts of bound ethereal solvent. Yttrocene and scandocene chloride complexes are the most common precursors for alkyl and hydride complexes, which tend to exhibit reactivity towards ethylene and α-olefins. A variety of alkylating reagents and reaction conditions were explored with little success. These complexes are well suited for formation of tetramethylaluminate complexes by reaction of either **3a** or **3b** with LiAlMe$_4$. Variable temperature ^{1}H NMR analysis suggests the monomeric nature of both $[(1,2\text{-SiMe}_2)_2\{\eta^5\text{-C}_5\text{H}_2\text{-4-(CMe}_3)\}\{\eta^5\text{-C}_5\text{H-3,5-(CHMe}_2)_2\}]\text{Y}(\mu\text{-Me})_2\text{AlMe}_2$ (**3c**) and $[(1,2\text{-SiMe}_2)_2\{\eta^5\text{-C}_5\text{H}_2\text{-4-(CMe}_3)\}\{\eta^5\text{-C}_5\text{H-3,5-(CHMe}_2)_2\}]\text{Sc}(\mu\text{-Me})_2\text{AlMe}_2$ (**3d**) (*16-17*). The reactivity of **3c** and **3d** with

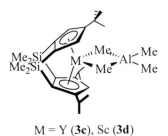

M = Y (**3c**), Sc (**3d**)

dihydrogen and α-olefins has been examined. While the yttrium complex appears to be unreactive at room temperature under an atmosphere of dihydrogen, the scandium complex reacts readily with dihydrogen to form a new species, as evidenced by ^{1}H NMR spectroscopy. Yttrocene **3c** requires added dihydrogen and heating to 87 °C to oligomerize α-olefins. In contrast, scandocene **3d** alone will oligomerize α-olefins such as 1-pentene and 1-hexene at room temperature without dihydrogen. Complex **3d** also serves as an effective hydrocyclization catalyst for α,ω-diolefins. At room temperature, **3d** reacts with 1,5-hexadiene in the presence of dihydrogen to form methylcyclopentane; similarly, 1,6-heptadiene is hydrocyclized to methylcyclohexane with catalyst **3d**. We are currently exploiting the reactivity of **3d** using isotopic perturbation of stereochemistry to probe for a transition state stabilizing α-agostic effect (*18-19*).

While this yttrocene tetramethylaluminate chemistry showed promise, efforts were directed towards generating a yttrocene hydride complex. Combination of **3a** and NaI in THF led to the formation of two new yttrocene halide species. Analysis by ^{1}H NMR spectral data suggests that one species is a dimeric iodide complex, while the other is an yttrium iodide/THF adduct. Further reaction of this yttrocene halide species with KCH(SiMe$_3$)$_2$ in toluene produces $((1,2\text{-SiMe}_2)_2\{\eta^5\text{-C}_5\text{H}_2\text{-4-(CMe}_3)\}\{\eta^5\text{-C}_5\text{H-3,5-(CHMe}_2)_2\})\text{YCH(SiMe}_3)_2$ (**3e**). Examination of the ^{1}H NMR spectral data reveals a major and minor component, which we attribute to two rotamers of the bis(trimethylsilyl)methyl group. Both components persist over a range of temperatures (196 K - 356 K). Addition of dihydrogen to **3e** leads to the formation of two new species; both appear to be dimeric yttrocene hydrides of the form $[((1,2\text{-SiMe}_2)_2\{\eta^5\text{-C}_5\text{H}_2\text{-4-(CMe}_3)\}\{\eta^5\text{-C}_5\text{H-3,5-(CHMe}_2)_2\})\text{YH}]_2$

(**3f**) (due to the diagnostic Y-H coupling observed in the ^{1}H NMR). While the x-ray crystal structure is not yet available for **3f**, we can postulate two possible dimeric structures. In addition, a "fly-over" dimeric structure (*20*) has not yet been ruled

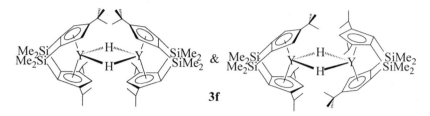

3f

out as a structural possibility. Combination of **3f** and various Lewis bases or olefins does not afford any new yttrocene complexes, but does perturb the relative amounts of the two dimers. Dilute solutions of olefins (ethylene or α-olefins) are consumed slowly and only upon heating in the presence of **3f**. In fact, successful polymerization of 1-pentene with this system requires in situ generation of **3f** by addition of dihydrogen to **3e** in neat 1-pentene. The polypentene formed appears to be atactic, with a slight preference for isotacticity, as determined by the ^{13}C NMR pentad analysis. For comparison, polypentene was also prepared using scandocene **3d** as the catalyst. This metallocene is far more active than the former system, yielding a viscous polymer after 1.5 hours at room temperature. The polypentene pentad analysis shows that it is atactic. The pentad analysis for both of these polypentene samples is listed in Table III. Preparation of polypropylene using catalysts **3f** and **3d** is currently in progress.

Table III: Pentad distribution (%) for polypentene produced by catalysts 3d and 3f.

catalyst	[*mmmm*]	[*rmmr*] [*mmmr*] [*mmmr*]	[*mmrm*] [*rmrr*]	[*mrmr*] [*rrrr*]	[*mrrr*]	[*mrrm*]
3d	6 %	26 %	28 %	20 %	18 %	5 %
3f	27 %	27 %	28 %	11 %	8 %	7 %

The formation of atactic polypentene with these group 3 systems can be rationalized by invoking a site epimerization mechanism. If site epimerization is facile and chain propagation is much slower with group 3 catalysts *vis-a-vis* cationic group 4 analogs, as is implied by their much lower activities, the metallocene will have the opportunity to pass through either diastereomeric transition state for olefin insertion. When site epimerization is faster than chain propagation, the regular alternation of olefin approach is disrupted and consequently syndiospecificity is

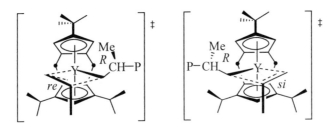

lost. The theoretical predictions for a pendant alkyl chain to rest in the center of the metallocene wedge can also be accommodated by this mechanism. In essence, facile swinging of the polymer chain prior to insertion is thwarting a syndiotactic preference for these C_s- symmetric group 3 metallocenes.

C_1- Symmetric Catalysts

C_1- symmetric catalysts **1d-e** exert a stereoselectivity which is more sensitive to propylene concentration than the analogous C_s- symmetric systems, **1a-c**. The stereospecificity dependence on propylene concentration, in terms of r-diad content, as well as that for $Me_2C(\eta^5-C_5H_3-3-Me)(\eta^5-C_{11}H_8)ZrCl_2$ (**2b**) is given in Figure 9.

With increasing propylene pressure, metallocenes **1d-e** evolve from iso- to syndiospecificity. In contrast, the stereocontrol for catalyst **2b** remains fairly constant over this concentration range. Since site epimerization is the primary error forming mechanism for catalysts **1a-c**, one can argue that the polymer microstructures produced by **1d-e** are a consequence of site epimerization. This is consistent with a unimolecular error forming mechanism competing with bimolecular chain propagation. At high propylene concentrations, regular migratory insertions are favored over site epimerization, producing polymer that is primarily syndiotactic. When propylene concentrations are decreased, site epimerization can effectively compete with chain propagation and syndiospecificity diminishes. If the bound polymer chain is given the opportunity to swing to the less sterically crowded side of the metallocene wedge prior to every olefin insertion, the resulting polymer will be primarily isotactic.

Since **2b** exhibits essentially no propylene concentration dependence on stereospecificity, it appears that it operates under a different stereocontrol mechanism than catalysts **1d-e**. This is consistent with Ewen's proposal for this system, with regularly alternating enchainment from stereospecific and aspecific sides of the metallocene wedge (*21-22*). Competitive site epimerization is not invoked in this system.

Despite these mechanistic differences, there are significant similarities in the polymer microstructures when catalysts **1d-e** and **2b** are utilized for propylene polymerization. Notably, when the r-diad content for polymers generated with **1d-**

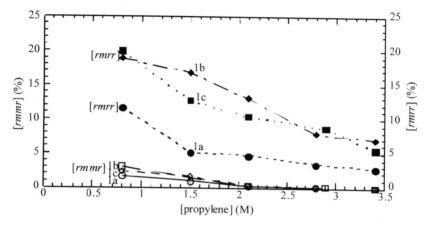

Figure 8. Dependence of *rmrr*- and *rmmr*- pentads on propylene concentration for C_s-symmetric catalysts **1a-c**/MAO.

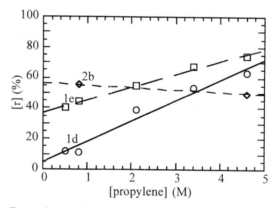

Figure 9. Dependence of *r*-diad content on propylene concentration for catalysts **1d-e** and **2b**/MAO.

e and **2b**/MAO is approximately 50 %, examination of the polymer microstructure reveals that the polymer is not atactic. The absence of the [*rmrm*] pentad and low intensity of [*rmrr*] + [*mrmm*] in addition to the relative pentad distribution of 3 : 2 : 1 : 4 : 0 : 0 : 3 : 2 : 1 (*mmmm : mmmr: rmmr : mmrr : rmrr + mrmm : rmrm : rrrr : rrrm : mrrm*) suggests a hemiisotactic polypropylene microstructure. While Ewen's stereocontrol mechanism provides a reasonable explanation for the hemiisotactic polymer microstructure produced by **2b**, the behavior of metallocenes **1d-e** can be accounted for by our site epimerization model. When a C_1- symmetric catalyst is used, site epimerization does not occur with the same rate in either direction. In fact, the sterics of the ligand array will provide "allowed" and "disallowed" chain swinging (Figure 10). Consideration of these "allowed" and "disallowed" pathways explains the absence of the [*rmrm*] pentad; this pentad results from enchainment achieved through a disallowed chain swing.

Examination of these C_1- symmetric systems has provided additional support for the importance of site epimerization in our doubly bridged metallocene systems. Further examination of the impact of ligand array substitution on stereoselectivity will be reported elsewhere (Veghini, D.; Henling, L. M.; Burkhardt, T. J.; Bercaw. J. E. *J. Am. Chem. Soc.,* submitted).

Conclusion

An easily modified doubly bridged ligand array has been utilized for the preparation of zirconocene, yttrocene, and scandocene complexes. These complexes serve as effective polymerization catalysts for α-olefins, with slight modifications in catalyst composition providing access to a variety of polymer microstructures. The polymer microstructure that results from using these C_s- and C_1- symmetric doubly-silylene bridged metallocenes can be attributed, in large part, to a site epimerization mechanism which perturbs syndioselectivity for a given system. Examination of polymer microstructure has provided a more complete understanding of the mechanism for producing syndiotactic poly(α-olefins).

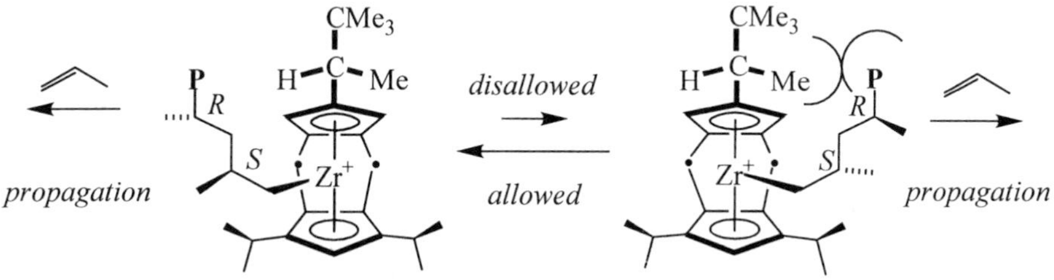

Figure 10. Site epimerization mechanism for an activated C_1- symmetric catalyst, **1d**.

14

Acknowledgments

The work had been supported by the Department of Energy (Grant No. DE-FG03-88ER13431) and by Exxon Chemicals America.

Literature Cited

(1)   Brintzinger, H. H.; Fischer, D.; Mülhaupt, R.; Rieger, B.; Waymouth, R. M. *Angew. Chem., Intl. Ed. Engl.* **1995**, *34*, 1143.

(2)   Kaminsky, W.; Kulper, K.; Brintzinger, H. H.; Wild, F. R. W. P. *Angew. Chem.* **1985**, *97*, 507; *Angew. Chem. Int. Ed. Engl.* **1985**, *24*, 507.

(3)   Ewen, J. A.; Elder, M. J. *Makromol. Chem., Macromol. Symp.* **1993**, *66*, 179.

(4)   Razavi, A.; Vereecke, D.; Peters, L.; Den Dauw, K.; Nafpliotis, L.; Atwood, J. L. In *Ziegler Catalysis, Recent Scientific Innovations and Technological Improvements*; Fink, F.; Mülhaupt, R.; Brintzinger, H. H., Eds.; Springer Verlag: Berlin, Heidelberg, 1995, p 111.

(5)   Corradini, P.; Guerra, G.; Vacatello, M.; Villani, V. *Gazz. Chim. Ital.* **1988**, *118*, 173.

(6)   Corradini, P.; Guerra, G.; Cavallo, L.; Moscardi, G.; Vacatello, M. In *Ziegler Catalysis, Recent Scientific Innovation and Technological Improvements*; Fink, G.; Mülhaupt, R.; Brintzinger H. H., Eds.; Springer-Verlag: Berlin, Heidelberg, 1995, p 237.

(7)   Guerra, G.; Corradini, P.; Cavallo, L.; Vacatello, M. *Makromol. Chem., Makromol. Symp.* **1995**, *89*, 77.

(8)   Gilchrist, J. H.; Bercaw, J. E. *J. Am. Chem. Soc.* **1996**, *118*, 12021.

(9)   Herzog, T. A.; Zubris, D. L.; Bercaw, J. E. *J. Am. Chem. Soc.* **1996**, *118*, 11988.

(10) Resconi, L.; Piemontesi, F.; Franciscono, G.; Abis, L.; Fiorani, T. *J. Am. Chem. Soc.* **1992**, *114*, 1025.

(11) Leclerc, M. K.; Brintzinger, H. H. *J. Am. Chem. Soc.* **1995**, *117*, 1651.

(12) Leclerc, M. K.; Brintzinger, H. H. *J. Am. Chem. Soc.* **1996**, *118*, 9024.

(13) Busico, V.; Cipullo, R. *J. Am. Chem. Soc.* **1994**, *116*, 9329.

(14) Busico, V.; Caporaso, L.; Cipullo, R.; Landriani, L. *J. Am. Chem. Soc.* **1996**, *118*, 2105.

(15) Bierwagen, E. P.; Bercaw, J. E.; Goddard, W. A. III *J. Am. Chem. Soc.* **1994**, *116*, 1481.

(16) Busch, M. A.; Harlow, R.; Watson, P. L. *Inorganica Chimica Acta* **1987**, *140*, 15.

(17) den Haan, K. H.; Wielstra, Y.; Eshuis, J. J. W.; Teuben, J. H. *J. Organomet. Chem.* **1987**, *323*, 181.

(18) Grubbs, R. H.; Coates, G. W. *Acc. Chem. Res.* **1996**, *29*, 85.

(19) Piers, W. E.; Bercaw, J. E. *J. Am. Chem. Soc.* **1990**, *112*, 9406.

(20) Stern, D.; Sabat, M.; Marks, T. J. *J. Am. Chem. Soc.* **1990**, *112*, 9558.

(21) Ewen, J. A.; Elder, M. J.; Jones, L.; Haspeslagh, L.; Atwood, J. L.; Bott, S. G.; Robinson, K. *Makromol. Chem., Makromol. Symp.* **1991**, *48/49*, 253.

(22) Herfert, N.; Fink, G. *Makromol. Chem., Makromol. Symp.* **1993**, *66*, 157.

Chapter 2

Recent Developments in Lanthanide Catalysts for 1,3-Diene Polymerization

L. Porri[1], G. Ricci[2], A. Giarrusso[1], N. Shubin[1,3], and Z. Lu[1,4]

[1]Department of Industrial Chemistry and Chemical Engineering,
Politecnico di Milano, Piazza Leonardo da Vinci 32, 20133 Milan, Italy
[2]Istituto di Chimica delle Macromolecole, CNR,
Via Bassini 15, 20133 Milan, Italy

Lanthanide catalysts are typical for the cis polymerization of 1,3-dienes. The ternary system AlEt$_2$Cl - Nd(OCOR)$_3$ - Al(iBu)$_3$ (R = alkyl group) is now used for the commercial production of cis polybutadiene. The effect of catalyst component addition order on activity and heterogeneity has been examined. In the ternary systems, only part of the lanthanide (Ln) is active in the catalysis. Much more active systems are obtained using allyl derivatives of Nd, Pr, Gd, in combination with methylaluminoxane (MAO). Evidence is reported suggesting that Ln catalysts have an ionic structure. Ln catalysts are also active for the polymerization of substituted butadienes and give cis-isotactic polymers from terminally substituted monomers.

Cis-polybutadiene (cis-PB) is a synthetic elastomer whose properties are very close to those of natural rubber. Since the beginning of the stereospecific polymerization various catalysts have been identified for the synthesis of this polymer (*1*). The first one, proposed in 1956, was based on iodine containing titanium compounds, TiI$_4$ or TiCl$_2$I$_2$, in combination with aluminum trialkyls, and gave a ca. 93% cis polymer (*2,3*). More stereospecific systems, capable of yielding a 97-98% cis-PB, were later obtained from cobalt (1957) (*4,5*) and nickel (1963) (*6-8*) compounds. In the first half of the 60s, some patents and papers appeared on the use of lanthanide catalysts for the cis polymerization of 1,3-dienes. Catalysts based on cerium (*9*) were the first used, but were soon abandoned, because the cerium residues catalyze the oxidation of the polymers. Catalysts based on neodymium, praseodymium and gadolinium were studied by Chinese groups (*10*). In Italy, Eni developed uranium systems, but no

---

[3]On leave from Institute of Macromolecular Compounds RAS, St. Petersburg, Russia.
[4]Current address: Department of Applied Chemistry, Tokyo University of Agriculture and Technology, 2–24–16 Nakacho, Koganei, Tokyo 184–8588, Japan.

practical application ensued, because of the problems connected with the uranium residues in the polymer (*1,11-14*).

The work carried out in the 60s and 70s showed that lanthanide catalysts have some advantages over the Ti, Co, Ni catalysts proposed before. In particular, they give a cis-PB more linear and hence more suitable for use in tires, which is by far the most important application (ca. 80%) for cis-PB. In addition, lanthanide catalysts are free of cationic activity and are more active in aliphatic than in aromatic solvents (*15*).

On the other hand, neodymium-based systems appeared as the most interesting among the lanthanide catalysts. They do not have the disadvantages of Ce or U systems and are more active than Gd or Pr systems. As a consequence, in recent years research work focused on Nd systems. This work has led to the identification of new catalytic compositions and to a better understanding of the nature of these systems. The present paper reports on the most significant results of this work.

Conventional Neodymium Systems

In their simplest form, Nd catalysts can be obtained by reacting a Nd compound [$Nd(acac)_3$; Nd-2-ethyl-hexanoate; $Nd(OCOC_7H_{15})_3$; $NdCl_3$; $NdCl_3$ complexed with donors] with an aluminum trialkyl. However, catalysts of practical interest are obtained only if Cl is present in the Nd compound. $Al(iBu)_3$-$Nd(OCOC_7H_{15})_3$ is a poor catalyst for butadiene polymerization and gives a polymer consisting of predominantly trans units (*16*). Catalyst activity and polymer cis content increase in the order: $Nd(OCOC_7H_{15})_3$ < $Nd(OCOC_7H_{15})_2Cl$ < $Nd(OCOC_7H_{15})Cl_2$ < $NdCl_3$. Among the aluminum alkyls, $Al(iBu)_3$ is more active than $AlEt_3$ and $AlMe_3$ (*17*). Various binary systems based on complexes $NdCl_3 \cdot nD$ (D = electron donor) and an aluminum alkyl are reported in patents and papers (*1,10,18*), but ternary systems based on a Nd carboxylate, a chlorine donor and $Al(iBu)_3$ are more useful commercially. $AlEt_2Cl$ or $Al_2Et_3Cl_3$ are the most commonly used chlorine donors, but tert-butyl chloride is reported in some patents and papers (*19*).

Two methods can be used for the preparation of the catalyst, differing for the order of catalyst component addition:

1) $Nd(OCOC_7H_{15})_3$ is first reacted with the chlorine donor, then with $Al(iBu)_3$ (*20*):

$$Nd(OCOC_7H_{15})_3 + 3AlEt_2Cl \rightarrow NdCl_3 + 3AlEt_2(OCOC_7H_{15}) \quad \text{(a)}$$
$$NdCl_3 + 30\ Al(iBu)_3 \rightarrow \text{active species} \quad \text{(b)}$$

2) $Nd(OCOC_7H_{15})_3$ is first reacted with $Al(iBu)_3$, then the chlorine donor is added (*19*):

$$Nd(OCOC_7H_{15})_3 + 30\ Al(iBu)_3 \rightarrow \text{Nd-Al complex} \quad \text{(c)}$$
$$\text{Nd-Al complex} + AlEt_2Cl \rightarrow \text{active species} \quad \text{(d)}$$

We have examined the influence of the order of catalyst component addition on both heterogeneity and activity of the system.

1. When $AlEt_2Cl$ is gradually introduced into a stirred heptane solution of $Nd(OCOC_7H_{15})_3$ at room temperature, a precipitate immediately forms, consisting of $NdCl_3$ or $NdCl_2(OCOC_7H_{15})$, depending on the Cl/Nd ratio. At the end of the

addition of AlEt$_2$Cl, stirring is continued for ca. 15 minutes, then Al(iBu)$_3$ is introduced (Al/Nd molar ratio about 30) into the stirred suspension. In our runs, the final Nd concentration was about $2 \div 4 \times 10^{-2}$ M. Reaction (a) is rapid and gives a finely subdivided precipitate consisting of the Nd-chloride. Reaction (b) is slow, as shown by the fact that the activity of the catalytic suspension increases with time, at room temperature (Figure 1; heterogeneous system).

2. The reaction between Nd(OCOC$_7$H$_{15}$)$_3$ and Al(iBu)$_3$ in heptane ([Nd] about $2 \cdot 10^{-2}$ M) does not produce a precipitate, at room temperature. Nothing is known about the product of this reaction, but it has been reported (21) that the reaction between AlMe$_3$ and Nd(O-^{t}Bu)$_3$ gives a bimetallic complex of structure [Nd(μ-O^{t}Bu)$_3$(μ-Me)$_3$(AlMe$_2$)$_3$]. It is plausible that analogous Al-Nd complexes are formed in reaction (c). However, this reaction is slow, as shown by the fact that addition of AlEt$_2$Cl gives different results depending on how long Nd(OCOC$_7$H$_{15}$)$_3$ and Al(iBu)$_3$ are let to react.

i) If AlEt$_2$Cl is added to the stirred solution of reaction (c) 1-2 hours after its beginning, a precipitate is slowly formed in a few hours; however, the precipitate appears after days for longer reaction times of (c).

ii) If reaction (c) is kept at room temperature for more than about 10 days, the subsequent addition of AlEt$_2$Cl does not produce a precipitate, even after weeks. A soluble catalyst is formed by this procedure.

However, the formation of the active species in reaction (d) is also a slow process. Figure 1 (homogeneous system) shows the variation of activity with time of a catalyst prepared by adding AlEt$_2$Cl to a solution of reaction (c) kept for ten days at room temperature. Time is measured starting from the addition of AlEt$_2$Cl.

Polymer cis content (> 96%) and intrinsic viscosity were found to be practically independent of catalyst aging time. The soluble catalysts gave polymers having molecular weights lower than the heterogeneous ones ([η] ca. 7.5 dL/g vs. ca. 10 dL/g).

Neodymium 2-ethyl-hexanoate and Nd-salts of versatic acids (branched fatty acids) are most commonly used for the catalyst preparation, as mentioned before. Some differences have been observed between the freshly prepared catalysts obtained from each compound, but the activity of the two systems tends to level off with aging.

Catalysts Based on Nd-allyl Compounds

The number of active centers for the heterogeneous system AlEt$_2$Cl-Nd(OCOC$_7$H$_{15}$)$_3$-Al(iBu)$_3$ has been determined by Russian authors and found to be rather low. About 7-8% of the Nd is active in freshly prepared catalysts (22). The data of Figure 1 indicate that aging the heterogeneous catalysts for about five weeks enhances significantly the activity, by a factor of about 3. The fact that polymer molecular weight is independent of catalyst aging time suggests that activity enhancement is mainly due to an increase in the number of active centers. However, the percentage of the metal active in catalysis (<25%) remains much lower than for other systems. For some metallocene catalysts, for instance, practically all the

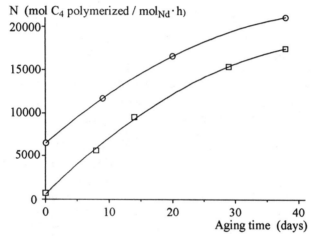

Figure 1. Effect of aging time on activity of Nd-systems.

○ AlEt$_2$Cl - Nd(OCOC$_7$H$_{15}$)$_3$ - Al(iBu)$_3$, hetereogeneous system;

□ Al(iBu)$_3$ - Nd(OCOC$_7$H$_{15}$)$_3$ - AlEt$_2$Cl, homogeneous system.

(Polymerization conditions: heptane, 16 mL; C$_4$H$_6$, 2 mL; Nd, 5·10^{-6} mol; room temperature).

transition metal has been reported to be active (23). The low number of active centers in Nd catalysts may be due to a slow alkylation of the Nd-compound and to the low stability of the Nd-alkyl bonds. These bonds are less stable than Nd-allyl bonds, which are formed on monomer insertion into the Nd-alkyl bonds .

If this interpretation is correct, Nd compounds containing Cl atoms and preformed Nd-allyl bonds should give, in combination with Al(iBu)$_3$, catalysts more active than the conventional ones. At the time we were working on this problem, the only known Nd compounds containing Nd-C bonds were Nd(benzyl)$_3$, stable only below -15°C (24), and Li[Nd(allyl)$_4$] (25). The latter compound was reported to have some activity, although rather low, for the polymerization of butadiene to trans polymers. It was decided to examine the reaction between Mg(allyl)Cl and NdCl$_3$ in THF as a possible route to chloro(allyl)derivatives of Nd.

Mg(allyl)Cl reacts smoothly with NdCl$_3$·2THF in THF, Mg/Nd molar ratio about 2-3, neodymium concentration about 0.04-0.1 M. We have first carried out the reaction at -20°C, but have later seen that it can be carried out at $0 \div +10$°C. After evaporation of THF in vacuo, a green powder, sensitive to air, was obtained (indicated as **A** in the following). Attempts to isolate well defined compounds by crystallizing the green powder from THF at -70°C gave only crystals of MgCl$_2$·2THF, while evaporation of THF after separation of the crystals gave an oily product, **B**, containing Nd and Mg in a molar ratio 1:1. Both **A** and **B** were used as the Nd compound for the preparation of the catalyst. The catalyst was usually prepared in toluene, where **A** is less soluble than **B**, at Al/Nd molar ratio 30, Nd concentration 0.01-0.02 M. The polymerization results showed that **A** and **B** give, in combination with various aluminum alkyls, catalysts much more active than the conventional ones (26, 27). Two years after our patent (26) was filed, Chinese workers also examined the reaction between Mg(allyl)Cl and NdCl$_3$ and succeeded in isolating a well defined complex by adding tetramethylethylenediamine (tmed) to the THF reaction solution. Well-shaped crystals of formula Nd(allyl)$_2$Cl·2MgCl$_2$·2tmed were obtained, whose structure was determined by X-ray (28). It was clear from the results of the Chinese authors that the compound obtained from Mg(allyl)Cl and NdCl$_3$ in THF was Nd(allyl)$_2$Cl·2MgCl$_2$·4THF.

Table I reports the results obtained using catalysts prepared from **A,** in combination with various aluminum alkyls, for the polymerization of butadiene. The relevant points are as follows.

1. **A** gives, in combination with Al(iBu)$_3$, catalysts more active, by a factor of about 3, than the conventional system AlEt$_2$Cl-Nd(OCOC$_7$H$_{15}$)$_3$-Al(iBu)$_3$. In addition, active catalysts are also obtained if AlMe$_3$ or AlEt$_3$ are used; it is known that these aluminum alkyls give poor catalysts in conventional systems (17).

2. Tetraisobutylaluminoxane, TIBAO, gives, in combination with **A**, catalysts more active than those obtained from Al(iBu)$_3$ or AlMe$_3$.

3. The most active catalysts are obtained using methylaluminoxane, MAO, as cocatalyst. In polymerizations carried out under the same conditions MAO-A has been found to be more active than AlEt$_2$Cl-Nd(OCOC$_7$H$_{15}$)$_3$-Al(iBu)$_3$ by a factor of about 30. Such an enhancement of the catalytic activity cannot depend only on an

increase in the number of active centers (7-8% of the Nd is active in the conventional system), but also on an increase in the kinetic constant of the propagation reaction.

4. Catalysts based on **A** give, independently of the aluminum alkyl used, polymers of lower molecular weight than those obtained with the conventional system $AlEt_2Cl-Nd(OCOC_7H_{15})_3-Al(iBu)_3$.

Table I. Polymerization of Butadiene with some Nd Catalysts [a]

Run	Catalyst [b]	Polymerization			$[\eta]^c$
		Time in min.	Conv. in %	N^d in h^{-1}	in $dL \cdot g^{-1}$
1	$AlEt_2Cl - Nd(OCOC_7H_{15})_3 - Al(iBu)_3$	15	23.5	4,800	> 10
2	$AlEt_2Cl - Nd(OCOC_7H_{15})_3 - AlMe_3$	120	<2	-	
3	$Nd(allyl)_2Cl \cdot 2MgCl_2 \cdot 4THF - Al(iBu)_3$	10	65.2	20,300	9.0
4	$Nd(allyl)_2Cl \cdot 2MgCl_2 \cdot 4THF - AlMe_3$	10	69.6	21,600	7.7
5	$Nd(allyl)_2Cl \cdot 2MgCl_2 \cdot 4THF - TIBAO$	3	57.8	60,000	6.0
6	$Nd(allyl)_2Cl \cdot 2MgCl_2 \cdot 4THF - MAO$	1.5	72.3	150,000	6.2

[a] Heptane 10 mL; Monomer 2 mL; Nd $5 \cdot 10^{-6}$ mol; 0°C;

[b] In runs 1 and 2, Cl/Nd/Al molar ratio 3/1/30; in runs 3-6, Nd/Al molar ratio 1/30

[c] Determined in toluene at 25°C.

[d] N = moles of monomer polymerized / $mol_{Nd} \cdot h$

5. Some aging time is necessary in order to reach maximum catalytic activity for the systems based on **A**. In the runs of Table I, catalysts have been aged at -20°C for 24 h, but aging conditions have not been optimized.

Catalysts prepared from **B** gave practically identical results.

Recently, German authors have succeeded in preparing $Nd(allyl)_3$, $Nd(allyl)_2Cl$ and $Nd(allyl)Cl_2$ via the following reactions:

$$Li[Nd(allyl)_4] + BEt_3 \rightarrow Li[BEt_3(allyl)] + Nd(allyl)_3$$
$$2Nd(allyl)_3 + NdCl_3 \rightarrow 3Nd(allyl)_2Cl$$
$$Nd(allyl)_2Cl + NdCl_3 \rightarrow 2Nd(allyl)Cl_2$$

They have reported on the activity of the catalysts derived from these complexes in combination with various Al alkyls (29-32). The most active system was found to be that based on MAO and $Nd(allyl)_2Cl$, which is the compound present in complexes **A** and **B**.

Praseodymium- and Gadolinium-based Catalysts

The ternary systems $AlEt_2Cl-Pr(OCOC_7H_{15})_3-Al(iBu)_3$ and $AlEt_2Cl-Gd(OCOC_7H_{15})_3-Al(iBu)_3$ have long been known and have been reported to be slightly less active than the corresponding Nd catalysts for the polymerization of dienes (*33*). We have reacted Mg(allyl)Cl with PrCl$_3$ and GdCl$_3$ in THF at 0÷+10°C (Mg/Pr or Gd molar ratio about 2-3) and have obtained in each case, after evaporation of THF in vacuo, a yellow microcrystalline powder, sensitive to air, corresponding to the formula Ln(allyl)$_2$Cl·2MgCl$_2$·4THF (Ln = Pr or Gd).

The Pr and the Gd allyl complexes, used in combination with MAO, gave catalysts having activity comparable to that of **A**-MAO. This indicates that also in the case of Gd and Pr catalysts the activity can be greatly enhanced by using bis-allyl derivatives of these compounds.

Nature of the Catalytic Species

Little is known about the structure of the active species in neodymium catalysts. However, the following findings seem to indicate that they have an ionic nature.

1. The polymerization of butadiene with the heterogeneous system AlEt$_2$Cl-Nd(OCOC$_7$H$_{15}$)$_3$-Al(iBu)$_3$ was found to be more rapid in CH$_2$Cl$_2$ than in heptane, by a factor of about 3 (Figure 2). The same has been found for the polymerization with the homogeneous system Nd(OCOC$_7$H$_{15}$)$_3$-Al(iBu)$_3$-AlEt$_2$Cl. It is to be noted that a similar phenomenon was observed for the polymerization of propene with iPr[FluCp]ZrCl$_2$-MAO and Me$_2$Si[Ind]$_2$ZrCl$_2$-MAO and was attributed to the ionic nature of the metallocene-MAO systems (*34*).

2. Complex **A** gives an active catalyst in combination with AlMe$_3$, as mentioned above. The activity of this system is, however, greatly enhanced by adding B(C$_6$F$_5$)$_3$ to the polymerization medium (Figure 3). Likely, B(C$_6$F$_5$)$_3$ reacts with AlMe$_3$, to produce AlMe$_2$(C$_6$F$_5$), which is a stronger Lewis acid than AlMe$_3$. An enhancement of activity due to an increase of the Lewis acidity of the cocatalyst speaks in favor of an ionic nature of the active species.

Polymerization of Substituted Butadienes

Neodymium systems are also active for the polymerization of various substituted butadienes (Table II). The relevant results can be summarized as follows.

Polymer Microstructure. All the monomers examined give polymers with a predominantly cis structure (*35-39*), with the exception of 2,4-hexadiene, which gives instead a trans polymer. This case will be examined separately. It is to be noted, however, that while some monomers (e.g., (E)-2-methyl-1,3-pentadiene; 2,3-dimethyl-1,3-butadiene) give, at room temperature, polymers consisting almost exclusively of cis units (>99%), other monomers give polymers with a lower cis content. (E)-1,3-pentadiene, for instance, gives, at room temperature, a polymer ca. 85% cis; the cis content increases by decreasing the polymerization temperature and is ca. 93% in the polymers obtained at -20°C. Isoprene and 2-ethyl-1,3-butadiene give polymers 93-94% cis at +20°C, the cis content increasing slightly by decreasing the polymerization temperature. In all the above polymers, with the exception of

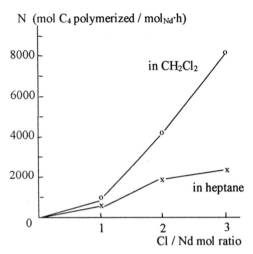

N (mol C₄ polymerized / mol_Nd·h)

Figure 2. Polymerization of butadiene with AlEt₂Cl - Nd(OCOR)₃ - Al(iBu)₃ in different solvents (Catalyst: Al/Nd = 30; [Nd] = $4.0 \cdot 10^{-2}$ M; aged at -20°C. Polymerization conditions: 0°C; [Nd] = $2.0 \cdot 10^{-4}$ M; monomer 20% v/v).

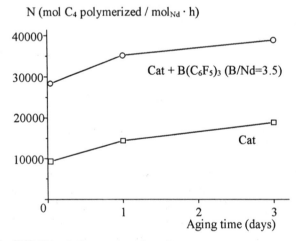

N (mol C₄ polymerized / mol_Nd · h)

Figure 3. B(C₆F₅)₃ influence on butadiene polymerization with allyl Nd catalysts. Cat = Nd(allyl)₂Cl·MgCl₂·nTHF - AlMe₃ in toluene, Al/Nd = 30; aged at -20°C. (Polymerization conditions: hexane, 45 mL; butadiene, 5 mL; Nd, $6 \cdot 10^{-6}$ mol; 25°C).

(E,E)-2,4-hexadiene as mentioned before, trans-1,4 units are practically absent, the non-cis units being 1,2 or 3,4.

Table II. Polymerization of substituted butadienes with AlEt$_2$Cl - Nd(OCOC$_7$H$_{15}$)$_3$ - Al(iBu)$_3$[a]

Monomer	N[b] (h^{-1})	Microstructure[c] cis-1,4 (%)	Microstructure[c] trans-1,4 (%)	$[\eta]$[d] (dL/g)	M.P. (°C)	I.P.[e] (Å)
Isoprene	1500	94		3.5		
2-Ethyl butadiene	1300	93		2.9		
(E)-1,3-Pentadiene	440	87		2.2	45	8.15
(E)-1,3-Hexadiene	400	85		1.3	85.6	8.0
(E)-3-Methyl pentadiene	300	80		2	68.9	8.02 6.97
2,3-Dimethyl butadiene	10	100			194.7	
(E)-2-Methyl pentadiene	2	100			160	7.9
(E,E)-2,4-Hexadiene	< 1		100		91.1	8.24

[a] Heptane 10 mL;   Monomer 2 mL;   Nd 1·10^{-5} mol;   0°C
[b] N = moles of monomer polymerized / mol$_{Nd}$ · h
[c] The remaining units are 1,2 (or 3,4)
[d] In toluene at 25 °C
[e] Identity period

It is well known that formation of a cis-1,4 versus a 1,2 (3,4) unit depends on whether the new monomer reacts at C1 or C3 of the last inserted unit (1,40,41). In the case of 2,3-dimethyl-1,3-butadiene, the formation of an exclusively cis polymer can be easily interpreted if one considers that, in this specific case, C3 is more substituted and, hence, much less reactive than C1. As a consequence, the new monomer will react only at C1, to give a practically 100% cis polymer (Figure 4).

For other monomers, such as isoprene, (E)-1,3-pentadiene, (E)-1,3-hexadiene, (E)-3-methyl-1,3-pentadiene, the formation of 1,2 or 3,4 units cannot be attributed only to a difference in reactivity between C1 and C3 af the allyl group. It is to be noted that a) these monomers can give, in principle, two types of allyl groups and b) they can react with one or the other enantioface.

The allyl groups that, in principle, these dienes can give on insertion into the growing chain are shown in Figure 5. Only allyl groups **a** are actually formed, as shown, for instance, by the fact that 1,2 isoprene units, –CH$_2$C(Me)(CH=CH$_2$)–, or 3,4 pentadiene units, –CH(Me)-CH(CH=CH$_2$)–, have never been observed in polyisoprenes and polypentadienes obtained with Nd catalysts (absence of vinyl group band at 911 cm^{-1} in the IR spectra). These units should be present, at least in very small amounts, if allyl groups **b** were formed.

Some structural features of the polymers obtained from isoprene, (E)-1,3-pentadiene, (E)-1,3-hexadiene and (E)-3-methyl-1,3-pentadiene give an

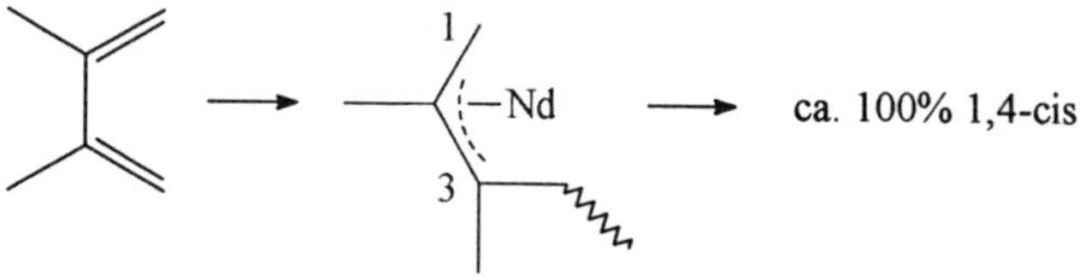

Figure 4. η^3-Butenyl group obtained from 2,3-dimethyl-1,3-butadiene on insertion into growing polymer chain.

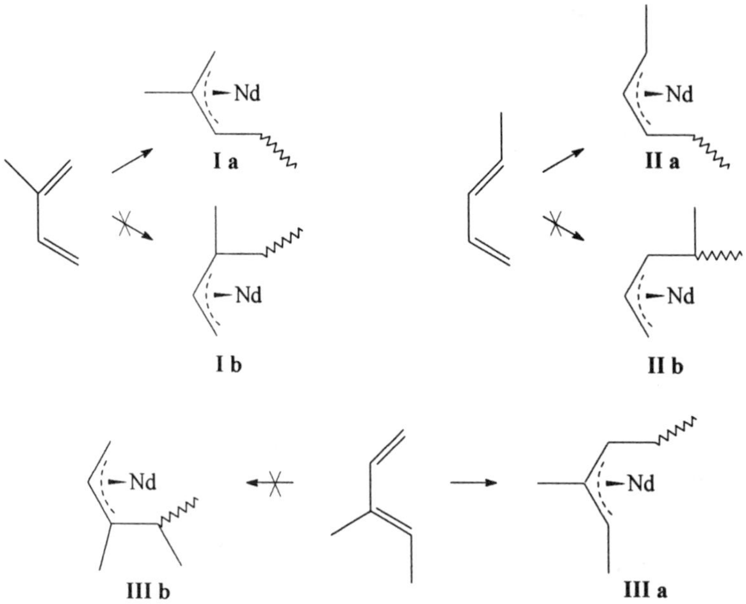

Figure 5. η^3-Butenyl groups that can be obtained from isoprene (I a, I b), (E)-1,3-pentadiene (II a, II b) and (E)-3-methyl-1,3-pentadiene (III a, III b) on insertion into the growing polymer chain (only the anti form is shown).

indication on how non-cis units are formed. The case of pentadiene is examined in Figure 6, as an example. Reaction of the monomer as in Figure 6a gives rise to a cis unit, while reaction of the monomer with the other enantioface, as in Figure 6b, will give a 1,2 unit since the allyl group **IIa** (Figure 5) is favored. A mechanism of this type should result in a polymer without inversions such as

$$-CH_2CH=CHCH(Me)-CH(Me)CH=CHCH_2-$$
and
$$-CH_2CH=CHCH(Me)-CH(CH=CHMe)CH_2-.$$

Such inversions have not been observed by NMR analysis, which supports the validity of the schemes proposed in Figure 6.

The same schemes, applied to the polymerization of 1,3-hexadiene or isoprene, account for the formation of, respectively, 1,2 and 3,4 units, along with cis units.

(E)-2-methyl-1,3-pentadiene also has two enantiofaces and it should behave as (E)-1,3-pentadiene. However, it gives polymers consisting almost exclusively of cis units. Our interpretation is that C3 of the allyl group derived from 2-methylpentadiene is practically unreactive, because more substituted than C1. Hence, presentation of the monomer as in Figure 7b does not result in an insertion, with consequent formation of cis units only.

Polymer Molecular Weight. Substituted butadienes give polymers with a molecular weight (MW) much lower than that of the PB obtained under the same conditions (Table II). The main process controlling polymer MW with Nd catalysts is the transfer reaction between the growing polymer chain and the aluminum alkyl, Al(iBu)$_3$ or Al(iBu)$_2$H (*10,42,43*):

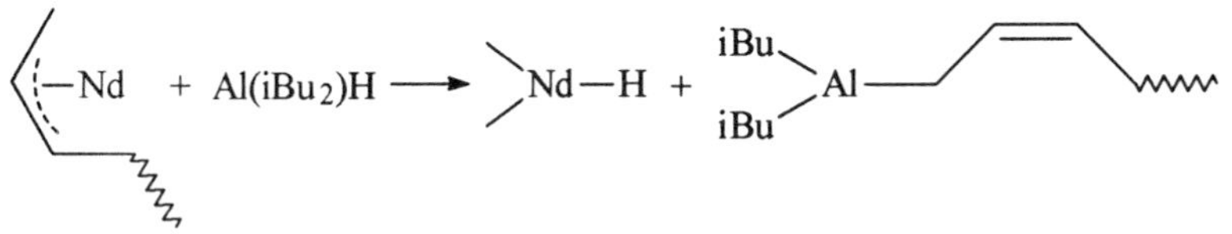

On the other hand, the polymerization of substituted butadienes is much slower than that of butadiene (Table II), indicating a lower value of Kp. Polymer MW is determined by the Kp/Kt ratio (Kp, Kt: kinetic constants for the propagation and termination reaction respectively). The values of these constants have not been determined, but the lower values of MW of poly(substituted butadienes) indicate that Kt does not decrease as much as Kp.

Polymers from Terminally Substituted Monomers. Terminally substituted monomers [(E)-1,3-pentadiene; (E)-2-methyl-1,3-pentadiene; (E)-3-methyl-1,3-pentadiene; (E)-1,3-hexadiene] give cis polymers having an isotactic structure.

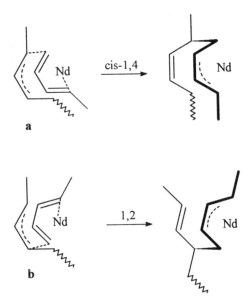

Figure 6. Scheme for the formation of a poly[(E)-1,3-pentadiene] with a mixed cis-1,4 / 1,2 structure. In **a** and **b** the new monomer is above the plane of the figure, the last inserted unit is below and Nd is on the plane.

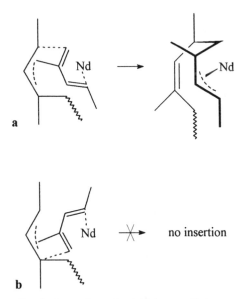

Figure 7. Scheme for the formation of poly[(E)-2-methyl-1,3-pentadiene] with a cis-1,4 isotactic structure. In **a** and **b** the new monomer is above the plane of the figure, the last inserted unit is below and Nd is on the plane.

Figure 8 shows the X-ray powder spectra of the polymers, while the identity periods are reported in Table II. Plausible schemes for the formation of isotactic polymers are shown in Figures 6a and 7a. With the orientation shown, the new monomer will give, after insertion, an allylic group of the same chirality as the previous one, as shown in previous papers by our group (*1,27,37*).

2,4-Hexadiene Polymers. The case of (E,E)-2,4-hexadiene has been treated in detail in previous papers (*44,45*). This monomer gives predominantly trans polymers even with catalysts that give cis polymers from common 1,3-dienes such as butadiene, isoprene and (E)-1,3-pentadiene. The formation of trans units is because this monomer polymerizes much more slowly than butadiene or isoprene, due to the steric hindrance of the methyl groups. The last inserted unit, which is *anti*-η^3-bonded to Nd, has time to isomerize to the more stable *syn* form, prior to monomer insertion. Reaction of the new monomer at C1 of the *syn* butenyl group gives rise to a trans unit. The validity of this interpretation has been confirmed by the results of the butadiene/2,4-hexadiene copolymerization using Nd catalysts. NMR analysis of the copolymers has shown that the hexadiene units that are isolated between butadiene units have a cis structure.

Conclusions

Lanthanide catalysts possess a set of properties that make them preferable to other catalysts, based on Ti, Co and Ni, for the cis polymerization of butadiene: a) they give more linear polymers, which are more suitable for use in tires; b) they are free of cationic activity; c) they are more active in aliphatic than in aromatic solvent.

The ternary systems Ln-carboxylate - chlorine donor - Al(iBu)$_3$ are the most known of this class of catalysts. Those based on Nd are currently used for the commercial production of cis-PB. The ternary systems can give soluble or insoluble catalytic species, depending on the order of catalyst component addition. Both soluble and insoluble catalysts give high-cis polybutadiene. Catalyst aging increases the activity, however the percentage of the lanthanide active in these systems is rather low, as compared with other organometallic polymerization catalysts. Much more active systems are obtained using allyl derivatives of the lanthanide, e.g., Ln(allyl)$_2$Cl, in combination with MAO.

Lanthanide catalysts give cis polymers also from various substituted butadienes. Stereoregular polymers with an isotactic structure are obtained from terminally substituted monomers.

Literature cited

1. Porri, L.; Giarrusso, A. In *Comprehensive Polymer Science;* Eastmond, G.C.; Ledwith, A.; Russo, S.; Sigwalt, P., Eds.; Pergamon Press, Oxford, UK, 1989, Vol.4, Part II; pp.53-108.
2. Zelinski, R.P.; Smith, D.R. (Phillips Petroleum Co.) Belgian Patent 551851 (Oct 17, 1956).
3. Phillips Petroleum Co., Br. Pat. 863337 (1961) - *Chem. Abstr.* **1961**, *55*, 20479.

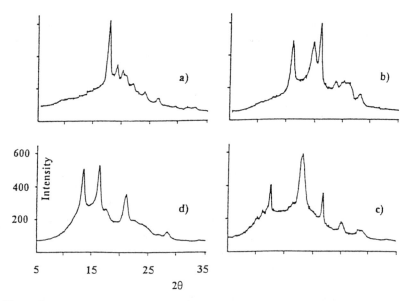

Figure 8. X-ray powder spectra of a) cis-1,4 isotactic poly[(E)-1,3-pentadiene]; b) cis-1,4 isotactic poly[(E)-1,3-hexadiene]; c) cis-1,4 isotactic poly[(E)-3-methyl-1,3-pentadiene]; d) cis-1,4 isotactic poly[(E)-2-methyl-1,3-pentadiene]. (Reproduced with permission from reference 46. Copyright 1998 Wiley.)

4. Longiave, C.; Croce, G.G.; Castelli, R. (Montecatini SpA) Italian Patent 592477 (Dec. 6, 1957).
5. Gippin, M. *Ind Eng. Chem, Prod. Res. Dev.* **1962**, *1*, 32.
6. Ueda, K.; Ohnishi, A.; Yoshimoto, T.; Hosono, J.; Maeda, K.; Matsumoto, T. *Kogyo Kagaku Zasshi*, **1963**, *66*, 1103.
7. Throckmorton, M.C.; Farson, F.S. *Rubber Chem. Technol.* **1972**, *45*, 268.
8. Saltman, W.; Kuzma, L.J. *Rubber Chem. Technol.* **1973**, *46*, 1055.
9. Throckmorton, M.C. *Kautsch. Gummi Kunstst.* **1969**, *22*, 293 - *Chem Abstr.* **1969**, *71*, 50995.
10. Hsieh, H.L.; Yeh, H.C. *Rubber Chem. Technol.* **1985**, *58*, 117.
11. Lugli, G.; Mazzei, A.; Poggio, S. *Makromol. Chem.* **1974**, *175*, 2021.
12. De Chirico, A.; Lanzani, P.C.; Raggi, E.; Bruzzone, M. *Makromol. Chem.* **1974**, *175*, 2029.
13. Bruzzone, M.; Mazzei, A.; Giuliani, G. *Rubber Chem. Technol.* **1974**, *47*, 1175.
14. Mazzei, A. *Makromol. Chem. Suppl.* **1981**, *4*, 61.
15. Ricci, G.; Boffa, G.; Porri, L. *Makromol. Chem., Rapid Commun.*, **1986**, *7*, 335.
16. Chigir, N.N.; Sharaev, O.K.; Tinyakova, E.I.; Dolgoplosk, B.A. *Vysokomol. Soedin.*, Ser. B, **1983**, *25*, 47.
17. Cabassi, F.; Italia, S.; Ricci, G.; Porri, L. In *Transition Metal Catalyzed Polymerizations;* Quirk, R.P., Ed.; Cambridge Univ. Press, MA, USA, 1988; pp.655-670.
18. Yang, J.-H; Tsutsui, M.; Chen, Z.; Bergbreiter, D.E. *Macromolecules* **1982**, *15*, 230.
19. Wilson, D.J.; Jenkins, D.K. *Polymer Bulletin* **1992**, *27*, 407.
20. Ricci, G.; Italia, S.; Cabassi, F.; Porri, L. *Polym. Commun.* **1987**, *28*, 223.
21. Biagini, B.; Lugli, G.; Abis, L.; Millini. R. *J. Organomet. Chem.* **1994**, *474*, C16.
22. Monakov, Y.B.; Marina, N.G.; Savele'va, I.G.; Zhiber, L.E.; Kozlov, V.G.; Rafikov, S.R. *Dokl. Akad. Nauk SSSR* **1982**, *265*, 1431.
23. Tait, P.J.T. In *Transition Metals and Organometallics as Catalysts for Olefin Polymerization*, Kaminsky, W. and Sinn, H., Eds.; Springer Verlag, Berlin, 1988; pp.309-327.
24. Chigir, N.N.; Guzman, I.Sh.; Sharaev, O.K.; Tiniakova, E.I.; Dolgoplosk, B.A. *Doklad. Akad. Nauk SSSR* **1982**, *265*, 1431.
25. Brunelli, M.; Poggio, S.; Pedretti, U.; Lugli, G. *Inorg. Chim. Acta* **1987**, *131*, 281.
26. Ricci, G.; Rotunno, D.; Italia, S.; Porri, L. Italian Patent 1228442, 19.06.1991 (It. priority 21 February 1989).
27. Porri, L.; Giarrusso, A.; Ricci, G. *Makromol. Chem., Macromol. Symp.* **1993**, *66*, 231.
28. Wu, W.; Chen, M.; Zhou, P. *Organometallics* **1991**, *10*, 98.
29. Taube, R.; Windisch, H.; Maiwald, S. *Macromol. Symp.* **1995**, *89*, 393.
30. Taube, R.; Maiwald, S.; Sieler, J. *J. Organomet. Chem.* **1996**, *513*, 37.
31. Taube, R.; Windisch, H.; Maiwald, S.; Hemling, H.; Schumann, H. *J. Organomet. Chem.* **1996**, *513*, 49.
32. Maiwald, S.; Taube, R.; Hemling, H.; Schumann, H. *J. Organomet. Chem.* **1998**, *552*, 195.
33. Shen, Z.; Ouyang, J.; Wang, F.; Hu, Z.; Yu, F.; Qian, B. *J. Polym. Sci., Polym. Chem. Ed.* **1980**, *18*, 3345.
34. Herfert, N.; Fink, G. *Makromol. Chem.* **1992**, *193*, 773.
35. Hsieh, H. L.; Heh, H. C. *Ind. Eng. Chem., Prod. Res. Dev.* **1986**, 25, 456.
36. Cabassi, F; Porzio, W.; Ricci, G.; Bruckner, S.; Meille, S. V.; Porri, L. *Makromol. Chem.* **1988**, 189, 2135.
37. Bolognesi, A.; Destri, S.; Porri, L.; Wang, F. *Makromol. Chem., Rapid Commun.* **1982**, 3, 187.
38. Ricci, G.; Zetta, L.; Meille, S. V. *Gazz. Chim. Ital.* **1996**, 126, 401.

39. Ricci, G.; Zetta, L.; Porri, L; Meille, S. V. *Macromol. Chem. Phys.* **1995**, 196, 2185.
40. Gallazzi, M.C.; Giarrusso, A.; Porri, L. *Makromol. Chem., Rapid Commun.* **1981**, 2, 59.
41. Porri, L.; Giarrusso, A.; Ricci, G. *Makromol. Chem., Macromol. Symp.* **1991**, 48/49, 239.
42. Witte, J. *Angew. Makromol. Chem.* **1981**, 94, 119.
43. Shen, Z.; Ouyang, J.; Wang, F.; Hu, Z.; Yu, F.; Qian, B. *J. Polym. Sci., Polym. Chem. Ed.* **1980**, 18, 3345.
44. Wang, F.; Bolognesi, A.; Immirzi, A.; Porri, L. *Makromol. Chem.* **1981**, 182, 3617.
45. Bolognesi, A.; Destri, S.; Zhou, Z.; Porri, L. *Makromol. Chem., Rapid Commun.* **1984**, 5, 679.
46. Porri, L.; Ricci, G.; Shubin, N. *Macromol. Symp.* **1998**, 128, 53.

Chapter 3

Siloxy-Substituted Group IV Metallocene Catalysts

Reko Leino and Hendrik J. G. Luttikhedde

Laboratory of Polymer Technology, Åbo Akademi University,
Biskopsgatan 8, FIN-20500, Åbo, Finland

The synthesis, characterization and olefin polymerization behavior of siloxy functionalized group IV bis(indenyl)-type metallocene catalysts is reviewed. Ethylene bridged 2- and 1-siloxy-substituted bis(indenyl) *ansa*-metallocenes are commonly obtained in moderate yields and high excess of the desired racemic diastereomer. In combination with methylaluminoxane (MAO) or other activators, the 2-substituted complexes form highly active catalyst systems for isospecific polymerization of propylene, homopolymerization of ethylene and copolymerization of ethylene with higher α-olefins. The 1-siloxy substituted catalysts are aspecific and show low activities in polymerization of propylene but show high activities in homo- and copolymerization of ethylene. The active catalysts can be generated with unusually low [Al]:[M] ratios (M = Zr, Hf), commonly ranging from (100-250):1 as compared with the conventional metallocene/MAO catalyst systems. The observations are attributed to the stabilizing and electron-donating effect from the sterically bulky Lewis-basic siloxy-substituent.

Metallocene catalysts for the polymerization of α-olefins are the focus of intense current interest.[1] Fourteen-electron cations[2] [L$_2$MR]$^+$ (L = cyclopentadienyl, indenyl, fluorenyl or related ligands; M = Ti, Zr, Hf; R = alkyl) have been identified as the active species, whereas activity and stereoregularity are determined by the steric and electronic properties of the ancillary ligand framework and the ion-ion interactions[3] between this highly electrophilic metallocene cation and its counterion.

Functionalization of metallocene catalysts with Lewis basic substituents would be desirable in order to tune the electronic properties of the catalyst precursors and to stabilize the cationic active sites essential for α-olefin coordination and chain propagation reaction. The chemistry of bis(cyclopentadienyl) complexes with pendant donor functionalized side chains has been reviewed recently.[4] Several disiloxane,[5] phosphine,[6] and pyridine[7] bridged group IV metallocenes have also been reported. The presence of N-Zr interactions in the latter complexes appears to favor the formation of zirconocene cations.[7c]

Halogen and methoxy substitution in the six-membered rings of the indenyl ligands decrease the catalytic activity and the polymer molecular weight.[8] These observations have been attributed to coordination of the donor groups to the methylaluminoxane (MAO) cocatalyst, resulting in inductive electron withdrawal. Similar effects have been reported for halogen and alkoxy substituted (fluorenyl)-(cyclopentadienyl) *ansa*-zirconocene/MAO catalysts,[9] and for alkoxy[10] and methylthio[11] substituted monoindenyltitanium trichloride/MAO catalyst systems.

Amino substituted bis(cyclopentadienyl)[12] and bis(indenyl)[13] complexes have been prepared in this and other laboratories. The bis(2-dimethylaminoindenyl) *ansa*-zirconocene/MAO catalyst systems show modest activities in polymerization of ethylene[13a] and propylene,[13c] despite of the unusually long induction periods.[13c] Other recent examples include phosphorus bridged[14] and propylthio substituted[15] bis-(cyclopentadienyl) and bis(indenyl) complexes.

1   **2**

We have recently reported the preparation of several siloxy substituted bis(indenyl) and bis(tetrahydroindenyl) metallocene complexes and their application in polymerization of α-olefins.[16,17] E.g., *rac*-[ethylenebis(2-(*tert*-butyldimethylsiloxy)-indenyl)]zirconium dichloride (**1**) and its 1-substituted analogue (**2**) are active catalyst precursors for homo- and copolymerization of ethylene. Complex **1**, in combination with MAO or other activators, is also highly active in isospecific polymerization of propylene. This paper reviews some of our earlier work on these complexes and summarizes the current status of the ongoing research.

Synthesis and Characterization of the Catalyst Precursors

Metallocene Synthesis. Generalized synthesis of the siloxy substituted metallocene complexes is presented in Scheme 1. Reaction of 2- or 1-indanone with *tert*-butyl-dimethylchlorosilane, thexyldimethylchlorosilane (thexyl = 1,1,2-trimethylpropyl) or tri-isopropylchlorosilane gives the corresponding 2- and 3-(trialkylsiloxy)indenes as distillable oils in 70-90% yield. Subsequent deprotonation with BuLi and the reaction of the lithium salt with 0.5 eq of dibromoethane gives the ethylene bridged ligand precursors as analytically pure crystalline solids in 45-70% yield, with the exception of

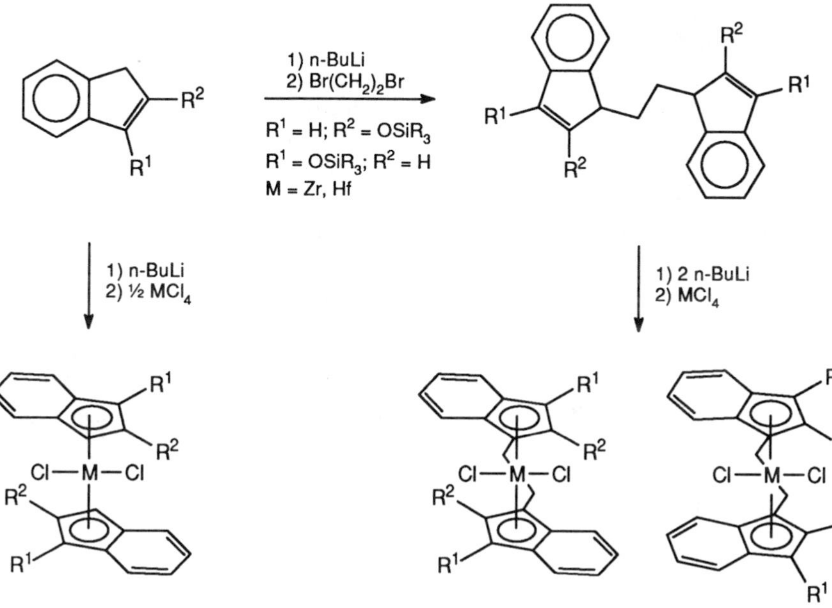

1) n-BuLi
2) Br(CH₂)₂Br

R¹ = H; R² = OSiR₃
R¹ = OSiR₃; R² = H
M = Zr, Hf

1) n-BuLi
2) ½ MCl₄

1) 2 n-BuLi
2) MCl₄

Scheme 1

bis(1-(triisopropylsiloxy)indenyl)ethane which is obtained as a fairly pure oil in nearly quantitative yield. Double deprotonation of the bridged ligands with BuLi and the subsequent reaction with ZrCl₄ gives the corresponding *ansa*-metallocene complexes in moderate 20-35% yields. The 2-siloxy substituted complexes are formed in a high excess of the racemic diastereomer, e.g., for complex **1** a 14:1 *rac/meso* ratio is observed, whereas for the bulkier 2-triisopropylsiloxy substituted analogue no *meso* form was detected by ^{1}H NMR analysis of the crude product. The pure racemic complexes are easily isolated by crystallization from an appropriate solvent. The 1-siloxy substituted complexes are formed as 5:1 mixtures of the *rac* and *meso* diastereomers. The racemic complex **2** has a better solubility in common organic solvents compared with its *meso* diastereomer and the pure *rac* form is obtained only after exhaustive recrystallizations. In the case of the 1-triisopropylsiloxy substituted analogue the *meso* diastereomer is more soluble and can be extracted from the crude mixture. The pure racemic complex can be isolated after crystallization. The corresponding 2- and 1-siloxy substituted bis(tetrahydroindenyl) complexes are obtained by catalytic hydrogenation of the parent metallocenes. A number of unbridged 1- and 2-siloxy substituted bis(indenyl) or bis(tetrahydroindenyl) complexes have also been prepared.

Separation of the *rac* and *meso* diastereomers of bis(indenyl) *ansa*-metallocene complexes is non-trivial since only the chiral C_2 symmetric racemic catalyst precursor produces isotactic polypropylene.[18] In many cases, separation from the undesired achiral *meso* isomer can not be achieved. The bulky siloxy substituent in the α-position to the ethylene bridge directs the metallation step in favor of the desired racemic diastereomer, presumably by steric interactions (e.g., complex **1**). In the β-position the effect is weaker but significant. The dimethylsilylene bridged analogue of **1** is, on the other hand, formed in an approximately 2:1 ratio of the *rac* and *meso* diastereomers and only small amounts of the fairly pure racemic complex have been obtained by fractional crystallization techniques.

Characterization. The siloxy substituted bis(indenyl) *ansa*-zirconocenes are generally soluble in cyclohexane, diethyl ether, toluene and dichloromethane. The hydrogenated congeners are even soluble in pentane and hexane. The ethylene bridged complexes show enhanced stabilities also in polar solvents, e.g., complex **1** can be recrystallized from acetone without decomposition. Colors of the 2- and 1-substituted complexes are remarkably different, apparently due to the different conjugation of the oxygen lone pairs with the indenyl moieties. Complex **1** is bright yellow and its hafnocene analogue light yellow, whereas complex **2** is bright orange and its hafnocene analogue bright yellow. The *meso* diastereomer of **1** is bright yellow and that of **2** dark red. All of the hydrogenated *ansa*-zirconocene complexes are light green/yellow.

Table I collects the ^{1}H NMR chemical shifts of the residual β-hydrogen atoms of various racemic 2-substituted bis(indenyl) and bis(tetrahydroindenyl) *ansa*-zirconocene dichlorides.[13a,13c,16a-c,19-25] The chemical shifts of the siloxy substituted complexes are shielded 0.6-0.7 ppm compared to the corresponding resonances of the 2-alkyl and unsubstituted congeners. Similar effects are observed in the ^{13}C NMR spectra for the chemical shifts of the β-carbon atoms and for the ^{1}H and ^{13}C NMR chemical shifts of the α-CH moieties of the 1-siloxy substituted bis(indenyl) and bis(tetrahydroindenyl) complexes as compared to those of the unsubstituted congeners (Table II).[16c,16e,18c,23-25] The observed shielding effects are consistent with increased electron density in the η^5-rings due to electron donation from the siloxy groups.

Table III shows the ^{29}Si and ^{13}C NMR chemical shifts for the Me_2Si moieties of complex 1, its hydrogenated congener and the corresponding ligand precursor. The p-d

Table I. ^1H NMR Chemical Shifts of the β-H Atoms of Various Racemic 2-Substituted Bis(indenyl) and Bis(tetrahydroindenyl) *ansa*-Zirconocene Dichlorides[a]

R	Et(2-RInd)₂ZrCl₂ β-H (ppm)[b]	Et(2-RIndH₄)₂ZrCl₂ β-H (ppm)[b]	Me₂Si(2-RInd)₂ZrCl₂ β-H (ppm)[b]	Ref.
*t*BuMe₂SiO	5.93	5.69	6.22	16a,b
thexMe₂SiO	5.95	-	-	16f
Me₂N	5.99	-	6.40	13a,c
*i*Pr₃SiO	6.03	5.75	-	16c
MeO	6.1-6.0	-	-	19
Ph	6.21	-	-	20
Me	6.65	6.06	6.80	21,22
Et	-	-	6.90	22
H	6.58	6.32	6.90	23-25

[a]β-H refers to the residual proton of the five-membered ring. [b]Downfield from TMS.

Table II. ^1H NMR Chemical Shifts of the α-H Atoms of Various Racemic 1-Substituted Bis(indenyl) and Bis(tetrahydroindenyl) *ansa*-Zirconocene Dichlorides[a]

R	Et(1-RInd)₂ZrCl₂ α-H (ppm)[b]	Et(1-RIndH₄)₂ZrCl₂ α-H (ppm)[b]	Me₂Si(1-Rind)₂ZrCl₂ α-H (ppm)[b]	ref.
*t*BuMe₂SiO	5.60	4.96	-	16e
*i*Pr₃SiO	5.63	5.01	-	16c
Me₃Si	6.10	-	-	18c
H	6.20	5.61	6.08	23-25

[a]α-H refers to the residual proton of the five-membered ring. [b]Downfield from TMS.

Table III. ^{29}Si and ^{13}C NMR Chemical Shifts for the Me_2Si Moieties of Complex 1, Its Ligand Precursor and the Hydrogenated Congener[a]

compound	^{29}Si (SiMe₂) (ppm)[b]	^{13}C (SiMe¹) (ppm)[b]	^{13}C (SiMe²) (ppm)[b]
Et(2-(*t*BuMe₂SiO)Ind)₂	22.45	-4.68	-4.88
rac-Et(2-(*t*BuMe₂SiO)Ind)₂ZrCl₂ (1)	26.64	-3.94	-4.09
rac-Et(2-(*t*BuMe₂SiO)IndH₄)₂ZrCl₂	25.21	-4.27	-4.65

[a]The ^{29}Si NMR spectra were recorded in CDCl₃ or CD₂Cl₂ solution on a JEOL JNM-LA400 NMR spectrometer. Gated decoupling was employed to suppress NOE. For the ^{13}C NMR data see ref. 16a and 16b. [b]Downfield from TMS.

conjugation between oxygen and silicon affects the electron density of the silicon atom in the ligand precursor, resulting in a higher field ^{29}Si resonance at 22.45 ppm. In the metallocene complexes, electron donation to the five-membered ring and the lewis acidic zirconium atom decreases the local electron density of the oxygen atom. Consequently, inductive electron withdrawal from silicon to oxygen results in lower field ^{29}Si resonances at 26.64 and 25.21 ppm respectively.

Molecular Structures. Table IV collects the selected bonding parameters for several siloxy substituted bis(indenyl) and bis(tetrahydroindenyl)metallocene dichlorides. All of the racemic *ansa*-zirconocenes show the expected C_2 symmetric bent metallocene structure with pseudotetrahedrally coordinated Zr atoms. The bond lengths and angles of the siloxy substituted complexes are well within the range observed previously for other ethylene-bridged bis(indenyl) and bis(tetrahydroindenyl)zirconium dichlorides.[13a,23,24,26-28] For the racemic bis(2-(siloxy)indenyl) complexes, the Zr-C bond lengths range from 2.476 Å to 2.625 Å, with the shortest distances involving the bridgehead carbon atoms. The Zr-C bonds of the hydrogenated congeners show slightly less variation ranging from 2.502-2.581 Å.

All of the racemic 2-siloxy substituted complexes, except *rac*-[ethylenebis(2-(triisopropylsiloxy)indenyl)]zirconium dichloride, crystallize in the less common indenyl-backward (Y) conformation, first described by Brintzinger for *rac*-Et(IndH$_4$)$_2$-Ti(O-acetyl-R-mandelate)$_2$,[29] and subsequently observed by Jordan for the [*rac*-Et(IndH$_4$)$_2$Zr(η^2-CH$_2$Ph)(CH$_3$CN)]$^+$ cation,[30] and recently for *rac*-Et(Ind)Hf(NMe)$_2$.[31] The solid state conformation of ethylene-bridged *ansa*-metallocenes is, at least to a certain extent, determined by intramolecular and intermolecular non-bonding contacts. Complexes with large equatorial substituents or ligands appear to favor the Y-conformation, although the energetical differences between the indenyl-backward and indenyl-forward (Π) conformations in the solid state are probably quite low. For the 2-siloxy-substituted complexes, the Y-conformation brings the chlorine and oxygen atoms in a closer contact (3.53-3.85 Å) than in the Π-conformation (4.06 Å). However, in the 1-siloxy substituted complexes even shorter Cl-O contacts are observed (3.15-3.30 Å), which are close to the sum of their relevant van der Waals radii (3.20 Å). Intuitively one could expect a stronger repulsion between the two electronegative atoms. On the other hand, electron donation from the oxygen to the five-membered ring and/or the zirconium atom may result in a sufficient electron deficiency that allows or enhances the short intramolecular distances. Similar hetero atom-chlorine close contacts were observed previously for the unbridged dimethylamino functionalized bis(cyclopentadienyl)[12] and bis(indenyl)[13a,13c] metallocene dichlorides.

Polymerization Catalysis

Polymerization of Propylene. Table V shows a comparison of several conventional and siloxy substituted *ansa*-zirconocene/MAO catalyst systems in polymerization of propylene. Under the employed conditions, the 2-*tert*-butyldimethylsiloxy substituted

Table IV. Selected Bonding Parameters for Siloxy Substituted Bis(indenyl) and Bis(tetrahydroindenyl)metallocene Dichlorides[a]

Zirconocene	Indenyl conformation[b]	M-Cen (av) (Å)	M-Cl (av) (Å)	ηC-O (av) (Å)	Cl-M-Cl (deg)	Cen-M-Cen (deg)	Cp-Cp (deg)	ηC-O-Si (av) (deg)	ref.
rac-Et(2-(*t*BuMe$_2$SiO)IndH$_4$)$_2$HfCl$_2$	backward	2.210	2.412	1.383	97.1	126.1	57.6	125.2	c
rac-Et(2-(*t*BuMe$_2$SiO)IndH$_4$)$_2$ZrCl$_2$	backward	2.231	2.439	1.376	98.2	125.2	58.6	125.1	16b
rac-Et(2-(*i*Pr$_3$SiO)IndH$_4$)$_2$ZrCl$_2$	backward	2.229	2.432	1.361	94.3	125.1	57.8	130.2	16c
meso-Et(2-(*t*BuMe$_2$SiO)IndH$_4$)$_2$ZrCl$_2$	staggered	2.232	2.439	1.361 d	97.1	125.5	57.9	d	c
rac-Et(2-(*t*BuMe$_2$SiO)Ind)$_2$ZrCl$_2$	backward	2.254	2.412	1.363	99.3	125.9	61.0	126.0	16a
rac-Et(2-(thexMe$_2$SiO)Ind)$_2$ZrCl$_2$	backward	2.243	2.433	1.356	97.9	120.3	58.6	129.0	16f
rac-Et(2-(*i*Pr$_3$SiO)Ind)$_2$ZrCl$_2$	forward	2.237	2.414	1.366	94.3	125.5	57.7	129.2	16c
rac-Et(1-(*t*BuMe$_2$SiO)IndH$_4$)$_2$ZrCl$_2$	forward	2.238	2.415	1.361	98.7	126.4	63.6	126.5	16e
meso-Et(1-(*t*BuMe$_2$SiO)Ind)$_2$ZrCl$_2$	staggered	2.252	2.431	1.341	96.2	126.4	64.6	133.0	16e
(2-(*t*BuMe$_2$SiO)-4,7-Me$_2$Ind)$_2$ZrCl$_2$	gauche	2.253	2.414	1.358	96.9	130.3	51.5	127.2	16d
(2-(*t*BuMe$_2$SiO)-4,7-Me$_2$Ind)$_2$ZrCl$_2$	syn	2.242	2.426	1.364	97.1	127.3	54.9	125.9	16d

[a]M = Zr, Hf; Cen refers to the centroids of the C$_5$ rings; (av) = average; Cp-Cp = angle between the cyclopentadienyl planes. [b]Conformation of the ligand backbone or the indenyl ligands. [c]Unpublished. [d]Disordered.

ethylene and dimethylsilylene bridged catalysts exhibit similar polymerization activities compared with their unsubstituted congeners and produce polypropylenes with higher melting points but lower molecular weights. The higher melting points are consistent with higher strereoregularities. The hydrogenated analogue of complex 1 shows a very low activity but produces at T_p = 20 °C isotactic polypropylene with higher molecular weight than the high-performance rac-Me$_2$Si(2-MeBenz[e]Ind)$_2$ZrCl$_2$/MAO catalyst. Its steroselectivity and the molecular weight of the produced polypropylene decline, however, rapidly with increasing polymerization temperature. For 1/MAO the polypropylene isotacticity declines from [mmmm] = 94-95% at T_p = 20 °C to the level of [mmmm] = 70% at T_p = 80 °C.

Table V. Comparison of Different Racemic ansa-Zirconocene/MAO Catalyst Systems in Polymerization of Propylene Under Similar Conditions[a]

Metallocene	A (kg PP/mol Zr/h)	M_w	M_w/M_n	T_m (°C)	ref.
rac-Me$_2$Si(2-MeBenz[e]Ind)$_2$ZrCl$_2$	7 200	98 500	2.1	150	16b
rac-Me$_2$Si(2-(tBuMe$_2$SiO)Ind)$_2$ZrCl$_2$	5 500	24 200	1.9	149	16f
rac-Et(Ind)$_2$ZrCl$_2$	5 400	27 900	2.2	131	16b
rac-Et(2-(tBuMe$_2$SiO)Ind)$_2$ZrCl$_2$ (1)	5 300	19 100	2.4	148	16a
rac-Me$_2$Si(Ind)$_2$ZrCl$_2$	4 400	54 200	1.8	142	16b
rac-Et(2-(thexMe$_2$SiO)Ind)$_2$ZrCl$_2$	2 700	16 100	2.1	146	16f
rac-C$_4$H$_8$Si(IndH$_4$)$_2$ZrCl$_2$	2 400	54 800	2.1	145	16b
rac-Me$_2$Si(IndH$_4$)$_2$ZrCl$_2$	2 300	53 200	2.0	146	16a
rac-Et(IndH$_4$)$_2$ZrCl$_2$	800	33 000	2.0	138	16b
rac-Et(2-(tBuMe$_2$SiO)IndH$_4$)$_2$ZrCl$_2$	20	123 200	2.1	150	16b
rac-Et(1-(tBuMe$_2$SiO)Ind)$_2$ZrCl$_2$ (2)	20	n.d.[b]	n.d.	n.d.	16e

[a]T_p = 20 °C; $P(C_3H_6)$ = 2.0 bar; [Al]:[Zr] = 3000:1; polymerization time = 60 min; [metallocene] = 11 μmol/200 mL of toluene. [b]Not determined.

The fairly low polypropylene molecular weights obtained with the siloxy substituted bis(indenyl) ansa-zirconocene/MAO catalyst systems result from extensive chain transfer to aluminum, which in the ^{13}C NMR spectra of the polymers is indicated by strong signals corresponding to isopropyl end groups. Similar observations were reported by Brintzinger and coworkers for the dimethylamino functionalized ansa-zirconocene catalysts.[13c] The exchange reaction should be enhanced with increased electron density at the metal center.[32]

The propylene polymerization activities of 1/MAO and its thexyl-substituted analogue show only little variation with cocatalyst:catalyst ratio between [Al]:[Zr] = (250-10000):1. The high activities at low [Al]:[Zr] ratios indicate a stable and easily generated active species, which we attribute to the stabilizing and electron-donating effect from the siloxy substituent. Conventional homogeneous metallocene/MAO catalyst systems require, in general, considerably higher ratios to achieve the maximum polymerization activities.[1a,18,22,27] [Al]:[Zr] ratios ranging from 2500:1 to 15000:1 are commonly reported in the literature. Decreased electrophilicity of the zirconium atom in the siloxy substituted complexes facilitates both the alkylation and ionization of the catalyst precursor and should increase the equilibrium amount of the solvent- or olefin separated cation, and enhance the rate of olefin insertion to the metal-carbon bond.

Polymerization of Ethylene. Figure 1 displays the ethylene polymerization activities of the MAO-activated catalyst precursors **1**, its hafnocene analogue rac-Et(2-(tBuMe$_2$SiO)Ind)$_2$HfCl$_2$, the hydrogenated congener rac-Et(2-(tBuMe$_2$SiO)IndH$_4$)$_2$-ZrCl$_2$ and the thexyl-analogue rac-Et(2-(thexMe$_2$SiO)Ind)$_2$ZrCl$_2$ as functions of the [Al]:[M] ratio (M = Zr, Hf). The bis(indenyl) complexes show maximum activities at unusually low [Al]:[M] ratios, ranging from 100:1 to 250:1, as observed earlier in polymerization of propylene. The tetrahydroindenyl zirconocene catalyst exhibits a different behavior with the ethylene polymerization activity increasing to a maximum value at [Al]:[Zr] = 3000:1. The siloxy-substituent clearly has a beneficial effect on the bis(indenyl) complexes, resulting in stable catalytic systems that can be activated with reduced amounts of the aluminoxane cocatalyst.

The siloxy substituted hafnium complex produces polyethylene with higher molecular weight than the correspondning zirconocene analogue. For all of the 2-siloxy substituted bis(indenyl) $ansa$-metallocene/MAO catalyst systems the molecular weights of the produced polyethylenes decrease strongly with increasing polymerization temperature and increasing [Al]:[M] ratio. The M_w of the polyethylene obtained with the hydrogenated congener of **1**/MAO exceeds 1 000 000 at T_p = 20-40 °C and is less dependent on the [Al]:[Zr] ratio. However, also for this catalyst system the M_w decreases drastically at higher polymerization temperatures.

Also the 1-siloxy substituted complexes, e.g., **2** and its tetrahydroindenyl analogue show high activities in polymerization of ethylene at reduced cocatalyst concentrations. Activity maxima are commonly observed at [Al]:[Zr] = (150-250):1. The activities are strongly dependent on the employed ratio and decline rapidly at high MAO-loadings. Molecular weights of the produced polyethylenes are generally lower than those obtained with 2-siloxy substituted analogues.

Both classes of the siloxy substituted complexes, i.e., the 2- and 1-substituted, form also highly active catalysts for copolymerization of ethylene with higher α-olefins. Detailed studies will be published elsewhere.[33]

Further Structural Variations

Background. Further structural variations of the siloxy-substituted complexes are required in order to improve their polymerization activities and the molecular weights and tacticities of the produced polymers. Based on the earlier studies by Spaleck,[18a,18c] Brintzinger[18b,18d] and Kaminsky[27] on structural modifications of 2-methyl substituted bis(indenyl) $ansa$-zirconocene catalysts, a similar incorporation of alkyl or aryl substituents to the six-membered rings of the siloxy substituted complexes could enhance their polymerization characteristics. E.g., Kaminsky and coworkers have reported very high stereoselectivities for the 2,4,7-trimethyl-substituted rac-Et(2,4,7-Me$_3$Ind)$_2$ZrCl$_2$/MAO catalyst system.[27] According to Brintzinger et $al.$, annelation of a

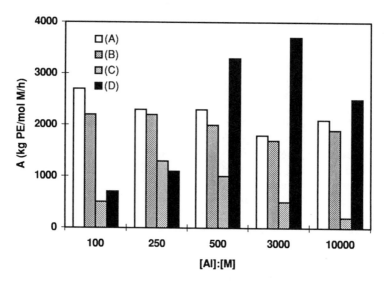

Figure 1. Ethylene polymerization activities of rac-Et(2-(tBuMe$_2$SiO)Ind)$_2$ZrCl$_2$ (**A**), rac-Et(2-(thexMe$_2$SiO)Ind)$_2$ZrCl$_2$ (**B**), rac-Et(2-(tBuMe$_2$SiO)Ind)$_2$HfCl$_2$ (**C**) and rac-Et(2-(tBuMe$_2$SiO)IndH$_4$)$_2$ZrCl$_2$ (**D**) as functions of the [Al]:[M] ratio (M = Zr, Hf) (T_p = 40 °C; P(C$_2$H$_4$) = 2.5 bar).

further aromatic six-membered ring to the 6,7-positions of *rac*-Me₂Si(2-MeBenz[*e*]-Ind)₂ZrCl₂ results in a six-fold increase in the propylene polymerization activity.[18d]

Metallocene Synthesis and Characterization. The preparation of 4-aryl/alkyl substituted 2-indanones is troublesome and commonly accomplished in only low overall yields after multi-step synthetic procedures.[22] However, 1,3-dihydro-2-oxocyclopenta-[*l*]phenanthrene (**3**) is readily available in good yield by reaction of phenanthrene-quinone with ethylacetoacetate.[34] Likewise, 4,7-dimethyl-2-indanone (**4**) is obtained in moderate yield by oxidation of 4,7-dimethylindene. Unfortunately, in both cases deprotonation of the corresponding 2-*tert*-butyldimethylsilyl enol ethers (**5** and **6**) results in nearly exclusive formation of the spiro(cyclopropane) adducts (**7** and **8**) (Scheme 2). The corresponding ethylene bridged ligand precursors were obtained in 4-5% yields under optimized reaction conditions. Formation of the cyclopropyl derivative by intramolecular bisalkylation is a major side reaction in the synthesis of ethylene bridged bis(indenyl) and bis(cyclopentadienyl) ligand precursors.[35] Its nearly exclusive domination is, however, less common. Apparently an unfavorable combination of steric and electronic properties enhances the rate of intramolecular alkylation of **5** and **6** instead of the desired intermolecular reaction.

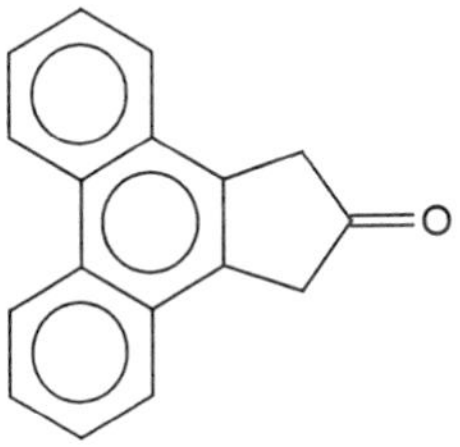

3 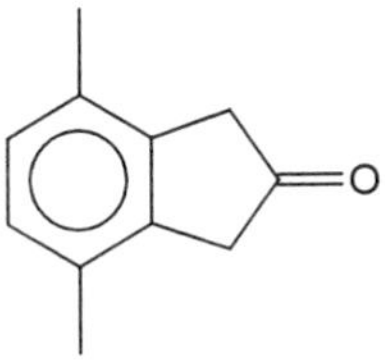   **4**

The unbridged zirconocene dichloride complexes **9** and **10** were obtained in 34% and 37% yields, respectively, by deprotonation of **5** resp. **6** and the subsequent reactions with ZrCl₄.

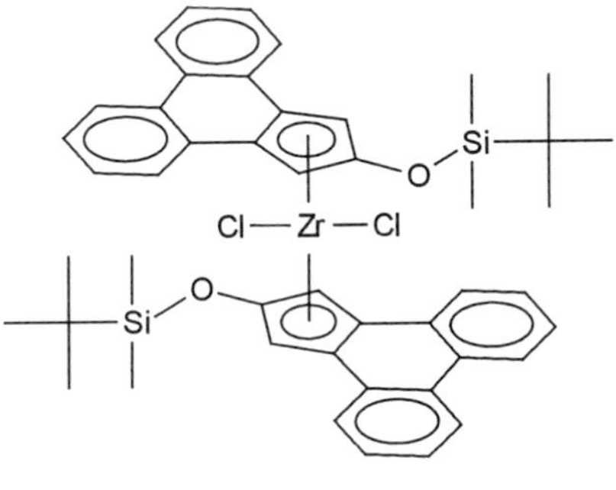

9 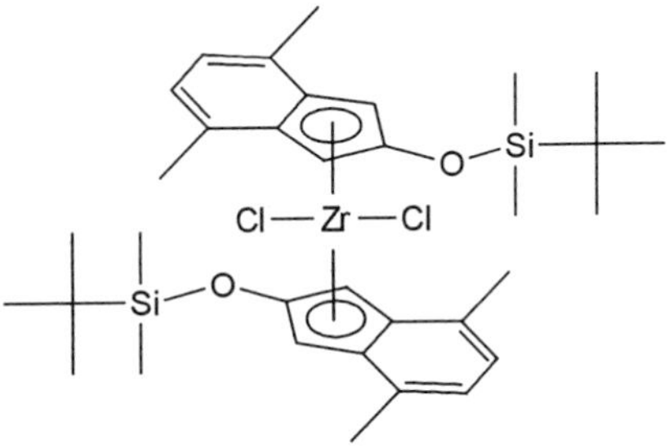   **10**

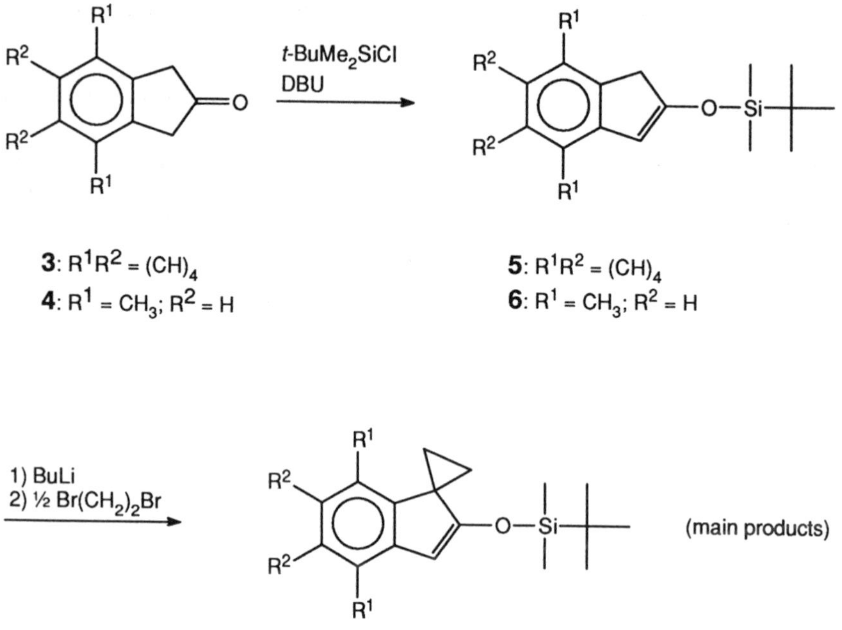

3: $R^1R^2 = (CH)_4$
4: $R^1 = CH_3$; $R^2 = H$

5: $R^1R^2 = (CH)_4$
6: $R^1 = CH_3$; $R^2 = H$

7: $R^1R^2 = (CH)_4$
8: $R^1 = CH_3$; $R^2 = H$

Scheme 2

X-ray structure determination of **10** revealed two rotamers in the asymmetric unit corresponding to a C_1 symmetric central/lateral:gauche conformation and a C_2 symmetric bis-central:syn conformation, respectively. In both rotamers, projections of the five-membered rings are nearly eclipsed instead of the commonly observed staggered gauche arrangement. For both complexes **9** and **10**, the room temperature ^{1}H and ^{13}C NMR data are consistent with time-averaged molecular C_{2v} symmetry.

The pure racemic dimethylsilylene bridged ligand **11** was obtained in 25% yield by reaction of three equivalents of **5-Li** with Me$_2$SiCl$_2$. The second equivalent of the anion reacts as a base deprotonating the phenanthrene moiety of the monochloro-silane.[18d] Addition of the third equivalent forms the bridged ligand. The *meso* diastereomer of **11** could not be isolated by fractional crystallization, although the initial crude product contained equal amounts of both isomers. Attempts to prepare the dimethylsilylene bridged ligand analogue of **6** failed yielding recovered **4** and **6** as the only isolated products. Double deprotonation of *rac*-**11** with BuLi and the subsequent reaction with ZrCl$_4$ gave a 1:9 mixture of *rac* and *meso*-[dimethylsilylenebis(2-(*tert*-butyldimethylsiloxy)cyclopenta[*l*]phenanthryl)]zirconium dichloride (**12**) (Scheme 3), as determined by ^{1}H NMR. Unfortunately, both diastereomers of **12** tend to decompose in common organic solvents. Pure *meso*-**12** was, however, obtained in 40% yield by crystallization from CH$_2$Cl$_2$.[36]

Polymerization Catalysis. Table VI collects selected ethylene polymerization results obtained with various zirconocene/MAO catalyst systems under similar conditions.[16a,16d,36-38] The unbridged complexes **9** and **10** show fairly low activities. For all siloxy-substituted catalyst systems higher molecular weights were obtained by decreasing the [Al]:[Zr] ratio, wheras the polymerization activity was practically independent of the employed cocatalyst concentration.

Table VI. Comparison of Different Zirconocene/MAO Catalyst Systems in Polymerization of Ethylene under Similar Conditions[a]

Metallocene	A (kg PE/ mol Zr/h)	M_w	M_w/M_n	T_m (°C)	ref.
Cp$_2$ZrCl$_2$	9 300	59 300	2.5	132	37
rac-Et(2-(*t*BuMe$_2$SiO)Ind)$_2$ZrCl$_2$ (**1**)	6 900	14 700	3.8	126	16a
meso-Et(2-(*t*BuMe$_2$SiO)Ind)$_2$ZrCl$_2$	4 900	21 000	2.6	124	38
meso-Me$_2$Si(2-(*t*BuMe$_2$SiO)Cp[*l*]Phen)$_2$ZrCl$_2$ (**12**)	4 400	17 000	2.2	n.d.[b]	36
(2-(*t*BuMe$_2$SiO)Cp[*l*]Phen)$_2$ZrCl$_2$ (**9**)	1 700	32 000	2.2	131	16d
(2-(*t*BuMe$_2$SiO)4,7-Me$_2$Ind)$_2$ZrCl$_2$ (**10**)	500	43 000	2.2	133	16d

[a]T_p = 80 °C; $P(C_2H_4)$ = 1.6 bar; [Al]:[Zr] = 3000:1; polymerization time = 20 min; [metallocene] = 11 µmol/200 mL of toluene. [b]Not determined.

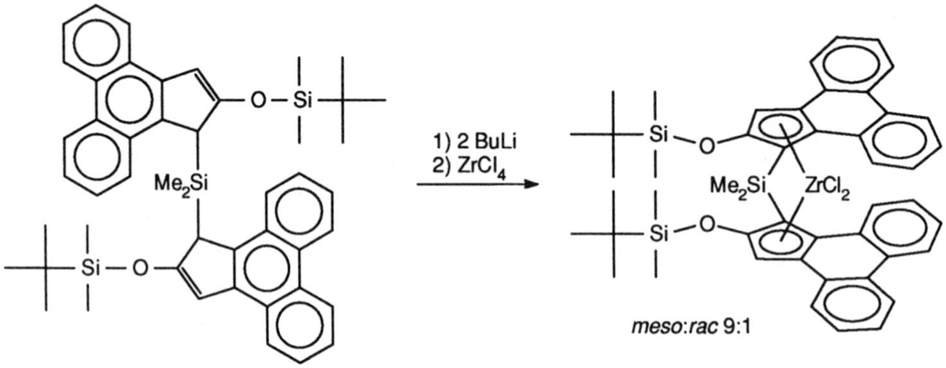

Scheme 3

Summary and Conclusions

Siloxy substituted group IV bis(indenyl) *ansa*-metallocene complexes form a new class of potential catalyst precursors for polymerization of α-olefins. In complex synthesis, the bulky substituent in α- or β-position to the interannular bridge directs the metallation step in favor of the desired racemic diastereomer. Pure racemic *ansa*-zirconocene complexes are in most cases obtained in moderate yields. The 2-siloxy substituted *ansa*-zirconocenes show, in combination with MAO or other activators, high activities in isospecific polymerization of propylene, homopolymerization of ethylene and copolymerization of ethylene with higher α-olefins. Very high polymerization activities are retained at exceptionally low [Al]:[Zr] ratios [(100-250):1]. The 1-siloxy substituted *ansa*-zirconocene catalysts are aspecific and practically inactive in polymerization of propylene but show high activities at low cocatalyst concentrations in homo- and copolymerization of ethylene. The presence of the interannular bridge appears to be required to achieve high activities. Most likely, free rotation of the bulky siloxy substituted ligands in the unbridged complexes forms considerable steric hindrance toward monomer coordination.

The following effects may contribute to the high polymerization activities at low [Al]:[Zr] ratios: (1) Decreased electrophilicity of the central metal which decreases the M-Cl bond strength and results in easier alkylation of the metallocene dichloride; (2) Weaker binding of the counterion to the cationic metallocene alkyl which increases the rate of monomer insertion to the metal-alkyl bond thus increasing the chain propagation rate; (3) Through-space or through-bonds electron donation from the Lewis-base functionalized ancillary ligand which stabilizes the cationic active site and favors its formation; (4) The bulky substituents favor the formation of active olefin separated ion pairs at the expense of dormant contact ion pairs.

Further synthetic and polymerization studies are in progress and will be reported in forthcoming publications.

Acknowledgments

Financial support from the Academy of Finland (Reko Leino and Jan H. Näsman, Siloxy Substituted Metallocenes), the Finnish Technology Development Center (TEKES) and Borealis Polymers is gratefully acknowledged. The authors wish to thank the other coauthors of the original publications for their contributions and Professor Jorma Mattinen for the ^{29}Si NMR measurements.

Literature Cited

1.   For recent reviews, see: (a) Brintzinger, H. H.; Fischer, D.; Mülhaupt, R.; Rieger, B.; Waymouth, R. M. *Angew. Chem., Int. Ed. Engl.* **1995**, *34*, 1143. (b) Bochmann, M. *J. Chem. Soc., Dalton Trans.* **1996**, 255.
2.   Jordan, R. F. *Adv. Organomet. Chem.* **1991**, *32*, 325.
3.   Piers, W. E. *Chem. Eur. J.* **1998**, *4*, 13, and references therein.
4.   Jutzi, P.; Redeker, T. *Eur. J. Inorg. Chem.* **1998**, *663*, and references therein.
5.   (a) Ciruelos, S.; Cuenca, T.; Gómez-Sal, P.; Manzanero, A.; Royo, P. *Organometallics* **1995**, *14*, 177. (b) Gräper, J.; Paolucci, G.; Fischer, R. D. *J.*

Organomet. Chem. **1995**, *501*, 211. (c) Naderer, H.; Siebel, E.; Fischer, R. D. *J. Organomet. Chem.* **1996**, *518*, 181. (d) Wang, B.; Su, L.; Xu, S.; Feng, R.; Zhou, X.; He, D. *Macromol. Chem. Phys.* **1997**, *198*, 3197.

6. Butchard, J. R.; Curnow, O. J.; Smail, S. J. *J. Organomet. Chem.* **1997**, *541*, 407.

7. (a) Qian, C.; Guo, J.; Ye, C.; Sun, J.; Zheng, P. *J. Chem. Soc., Dalton Trans.* **1993**, 3441. (b) Peng, K.; Xiao, S. *Makromol. Chem., Rapid Commun.* **1993**, *14*, 633. (c) Thiele, K.-H.; Schließburg, C.; Neumüller, B. *Z. Anorg. Allg. Chem.* **1995**, *621*, 1106. (d) Paolucci, G.; Pojana, G.; Zanon, J.; Lucchini, V.; Avtomonov, E. *Organometallics* **1997**, *16*, 5312.

8. (a) Piccolrovazzi, N.; Pino, P.; Consiglio, G.; Sironi, A.; Moret, M. *Organometallics* **1990**, *9*, 3098. (b) Lee, I.-M.; Gauthier, W. J.; Ball, J. M.; Iyengar, B.; Collins, S. *Organometallics* **1992**, *11*, 2115.

9. (a) Ewen, J. A. *Macromol. Symp.* **1995**, *89*, 181. (b) Alt, H. G.; Zenk, R. *J. Organomet. Chem.* **1996**, *522*, 39.

10. Foster, P.; Rausch, M. D.; Chien, J. C. W. *J. Organomet. Chem.* **1997**, *527*, 71.

11. Ready, T. E.; Chien, J. C. W.; Rausch, M. D. *J. Organomet. Chem.* **1996**, *519*, 21.

12. Stahl, K.-P.; Boche, G.; Massa, W. *J. Organomet. Chem.* **1984**, *277*, 113.

13. (a) Luttikhedde, H. J. G.; Leino, R. P.; Wilén, C.-E.; Näsman, J. H.; Ahlgrén, M. J.; Pakkanen, T. A. *Organometallics* **1996**, *15*, 3092. (b) Plenio, H.; Burth, D. *J. Organomet. Chem.* **1996**, *519*, 269. (c) Barsties, E.; Schaible, S.; Prosenc, M.-H.; Rief, U.; Röll, W.; Weyand, O.; Dorer, B.; Brintzinger, H.-H. *J. Organomet. Chem.* **1996**, *520*, 63. (d) Luttikhedde, H. J. G.; Leino, R. P.; Ahlgrén, M. J.; Pakkanen, T. A.; Näsman, J. H. *J. Organomet. Chem.* **1998**, *557*, 225.

14. (a) Leyser, N.; Schmidt, K.; Brintzinger, H.-H. *Organometallics* **1998**, *17*, 2155. (b) Schaverien, C. J.; Ernst, R.; Terlouw, W.; Schut, P.; Sudmeijer, O.; Budzelaar, P. H. M. *J. Mol. Catal. A: Chem.* **1998**, *128*, 245.

15. Broussier, R.; Bourdon, C.; Blacque, O.; Vallat, A.; Kubicki, M. M.; Gautheron, B. *J. Organomet. Chem.* **1997**, *538*, 83.

16. (a) Leino, R.; Luttikhedde, H.; Wilén, C.-E.; Sillanpää, R.; Näsman, J. H. *Organometallics* **1996**, *15*, 2450. (b) Leino, R.; Luttikhedde, H. J. G.; Lehmus, P.; Wilén, C.-E.; Sjöholm, R.; Lehtonen, A.; Seppälä, J. V.; Näsman, J. H. *Macromolecules* **1997**, *30*, 3477. (c) Luttikhedde, H. J. G.; Leino, R.; Lehtonen, A.; Näsman, J. H. *J. Organomet. Chem.* **1998**, *555*, 127. (d) Leino, R.; Luttikhedde, H. J. G.; Lehtonen, A.; Sillanpää, R.; Penninkangas, A.; Strandén, J.; Mattinen, J.; Näsman, J. H. *J. Organomet. Chem.* **1998**, *558*, 171. (e) Leino, R.; Luttikhedde, H. J. G.; Lehtonen, A.; Ekholm, P.; Näsman, J. H. *J. Organomet. Chem.* **1998**, *558*, 181. (f) Leino, R.; Luttikhedde, H. J. G.; Lehmus, P.; Wilén, C.-E.; Sjöholm, R.; Lehtonen, A.; Seppälä, J. V.; Näsman, J. H. *J. Organomet. Chem.* **1998**, *559*, 65.

17. Related work on siloxy substituted bis(cyclopentadienyl), monocyclopentadienyl and monoindenyl group IV metallocene complexes has appeared recently, see: (a) Plenio, H.; Warnecke, A. *J. Organomet. Chem.* **1997**, *544*, 133. (b) Tian, G.; Xu, S.; Zhang, Y.; Wang, B.; Zhou, X. *J. Organomet. Chem.* **1998**, *558*, 231. In addition, similar constrained geometry-type complexes have been reported in the patent literature, see: (c) Klosin, J.; Kruper, W. J., Jr.; Nickias, P. N.; Patton, J. T.; Wilson, D. R. PCT Int. Appl. WO 98/06728, 1998.

18. See, for example: (a) Spaleck, W.; Küber, F.; Winter, A.; Rohrmann, J.; Bachmann, B.; Antberg, M.; Dolle, V.; Paulus, E. F. *Organometallics* **1994**, *13*, 954. (b) Stehling, U.; Diebold, J.; Kirsten, R.; Röll, W.; Brintzinger, H.-H.; Jüngling, S.; Mülhaupt, R.; Langhauser, F. *Organometallics* **1994**, *13*, 964. (c) Spaleck, W.; Antberg, M.; Aulbach, M.; Bachmann, B.; Dolle, V.; Haftka, S.; Küber, F.; Rohrmann, J.; Winter, A. In *Ziegler Catalysts*; Fink, G.; Mülhaupt, R.; Brintzinger, H. H., Eds.; Springer-Verlag: Berlin, Heidelberg, 1995; p 83. (d)

Schneider, N.; Huttenloch, M. E.; Stehling, U.; Kirsten, R.; Schaper, F.; Brintzinger, H. H. *Organometallics* **1997**, *16*, 3413.

19. Winter, A.; Rohrmann, J.; Antberg, M.; Spaleck, W.; Herrmann, W. A.; Riepl, H., Eur. Pat. Appl. EP 582 195, 1994.

20. Waymouth, R. M.; Hauptman, E.; Coates, G. W. PCT Int. Appl. WO 95/25757, 1995.

21. Winter, A.; Antberg, M.; Spaleck, W.; Rohrmann, J.; Dolle, V. US Patent 5 276 208, 1994.

22. Spaleck, W.; Antberg, M.; Rohrmann, J.; Winter, A.; Bachmann, B.; Kiprof, P.; Behm, J.; Herrmann, W. A. *Angew. Chem., Int. Ed. Engl.* **1992**, *31*, 1347.

23. Piemontesi, F.; Camurati, I.; Resconi, L.; Balboni, D.; Sironi, A.; Moret, M.; Zeigler, R.; Piccolrovazzi, N. *Organometallics* **1995**, *14*, 1256.

24. Collins, S.; Kuntz, B. A.; Taylor, N. J.; Ward, D. G. *J. Organomet. Chem.* **1988**, *342*, 21.

25. Herrmann, W. A.; Rohrmann, J.; Herdtweck, E.; Spaleck, W.; Winter, A. *Angew. Chem., Int. Ed. Engl.* **1989**, *28*, 1511.

26. Ewen, J. A.; Elder, M. J.; Jones, R. L.; Haspeslagh, L.; Atwood, J. L.; Bott, S. G.; Robinson, K. *Makromol. Chem., Macromol. Symp.* **1991**, *48/49*, 253.

27. Kaminsky, W.; Rabe, O.; Schauwienold, A.-M.; Schupfner, G. U.; Hanss, J.; Kopf, J. *J. Organomet. Chem.* **1995**, *497*, 181.

28. Resconi, L.; Piemontesi, F.; Camurati, I.; Balboni, D.; Sironi, A.; Moret, M.; Rychlicki, H.; Zeigler, R. *Organometallics* **1996**, *15*, 5046.

29. (a) Schäfer, A.; Karl, E.; Zsolnai, L.; Huttner, G.; Brintzinger, H.-H. *J. Organomet. Chem.* **1987**, *328*, 87. (b) Brintzinger, H. H. In *Transition Metals and Organometallics as Catalysts for Olefin Polymerization*; Kaminsky, W.; Sinn, H.-J., Eds.; Springer-Verlag: Berlin, 1988; p 249.

30. Jordan, R. F.; LaPointe, R. E.; Baenziger, N.; Hinch, G. D. *Organometallics* **1990**, *9*, 1539.

31. Diamond, G. M.; Jordan, R. F.; Petersen, J. L. *Organometallics* **1996**, *15*, 4030.

32. Siedle, A. R.; Newmark, R. A.; Lamanna, W. M.; Schroepfer, J. N. *Polyhedron* **1990**, *9*, 301.

33. (a) Härkki, O.; Lehmus, P.; Leino, R.; Luttikhedde, H. J. G.; Näsman, J. H.; Seppälä, J. V., submitted to *Macromol. Chem. Phys.* (b) Lehmus, P.; Kokko, E.; Härkki, O.; Leino, R.; Luttikhedde, H. J. G.; Näsman, J. H.; Seppälä, J. V., manuscript.

34. (a) Japp, F. R.; Streatfeild, F. W. *J. Chem. Soc.* **1883**, *43*, 27. See also: (b) Cope, A. C.; Field, L.; MacDowell, D. W. H.; Wright, M. E. *J. Am. Chem. Soc.* **1949**, *71*, 1589.

35. Halterman, R. L. *Chem. Rev.* **1992**, *92*, 965.

36. For experimental details concerning the preparation of the cyclopenta[*l*]-phenanthryl derivatives see: Luttikhedde, H. J. G.; Leino, R.; Wilén, C.-E.; Näsman, J. H. *Polym. Prepr. (Am. Chem. Soc., Div. Polym. Chem.)* **1998**, *39(1)*, 229.

37. Wilén, C.-E.; Luttikhedde, H.; Hjertberg, T.; Näsman, J. H. *Macromolecules* **1996**, *29*, 8569.

38. Leino, R.; Luttikhedde, H. J. G.; Lehtonen, A.; Strandén, J.; Mattinen, J.; Näsman, J. H. *Polym. Prepr. (Am. Chem. Soc., Div. Polym. Chem.)* **1998**, *39(1)*, 228.

PROGRESS IN POLYMERIZATION

Chapter 4

Advances in Propene Polymerization Using Magnesium Chloride-Supported Catalysts

J. C. Chadwick[1], G. Morini[1], G. Balbontin[1], V. Busico[2], G. Talarico[2], and O. Sudmeijer[3]

[1]Montell Polyolefins, Centro Ricerche G. Natta, Piazzale Donegani 12, 44100 Ferrara, Italy
[2]Università di Napoli "Federico II", Via Mezzocannone 4, 80134 Napoli, Italy
[3]Shell Research and Technology Centre Amsterdam, 1030 BN Amsterdam, the Netherlands

Studies on the effects of hydrogen, monomer concentration and temperature in propene polymerization using high-activity MgCl$_2$-supported catalysts have revealed that the effects of the catalyst on polymer yield, isotacticity and molecular weight are interrelated and dependent on both regio- and stereoselectivity. The high hydrogen sensitivity of catalysts containing a diether donor is mainly due to chain transfer following the occasional regioirregular (2,1-) monomer insertion. The probability of regio- and stereoirregular insertion decreases with monomer concentration, indicating the presence of non-symmetric active centres. The stereoregularity of the isotactic polymer also increases with polymerization temperature, indicating a greater relative increase in productivity for highly isospecific as opposed to moderately isospecific centres.

High-activity magnesium chloride-supported Ziegler-Natta catalysts play a dominant role in polypropylene (PP) manufacture. Current worldwide manufacturing capacity for PP exceeds 20 million tonnes per annum and most of this utilizes catalysts comprising MgCl$_2$, TiCl$_4$ and an electron donor. Three different types of Lewis base have been found to be effective as "internal" donor in the catalyst: monoesters (*1*) (e.g. ethyl benzoate), diesters (*2*) (e.g. diisobutyl phthalate) and diethers (*3,4*) (typically 2,2-disubstituted-1,3-dimethoxypropanes). The cocatalyst used in polymerization is a trialkylaluminium such as AlEt$_3$ and for high stereoselectivity the ester-containing catalysts normally require the presence in polymerization of an additional Lewis base, termed external donor. This is typically a second aromatic ester for catalysts containing ethyl benzoate, and an alkoxysilane for catalysts containing a phthalate ester. The recently-developed catalysts containing a diether as internal donor do not require an external donor for high stereoselectivity, as the diether has a higher affinity towards $MgCl_2$ than towards AlR_3 and is therefore, in contrast to the esters, not displaced on contact with the cocatalyst (5).

The performance of $MgCl_2$-supported catalysts in propene polymerization is strongly dependent on the donor(s) present in the catalyst system and on process conditions such as the presence of hydrogen, the monomer concentration and temperature. At the moment the most widely used catalyst system is $MgCl_2/TiCl_4$/diisobutyl phthalate - AlR_3 - alkoxysilane. The structure of the alkoxysilane, typically $RR'Si(OMe)_2$, affects not only the isotacticity of the polymer but also the polypropylene yield, molecular weight and molecular weight distribution. The newly-developed and highly active catalyst system $MgCl_2/TiCl_4$/diether - AlR_3 gives particularly high polymer yields, while an added advantage of this system is that relatively little hydrogen is required for molecular weight control.

The purpose of this paper is to pinpoint the fundamental factors determining the performance of state-of-the-art Ziegler-Natta catalysts for polypropylene, our recent studies having demonstrated that both stereo- and regioselectivity play an important role in determining not only the polymer isotacticity but also the catalyst activity and sensitivity to hydrogen.

Experimental

Catalyst Preparation and Polymerization. Catalyst preparation was carried out as described previously (6-8). Polymerizations in liquid propene were carried out at 67 or 70 °C (6,8), using $AlEt_3$ as cocatalyst. Polymerizations in heptane slurry were carried out at 50 °C, keeping the monomer pressure constant (9).

Polymer Analysis. Xylene-insoluble fractions were isolated by dissolving 2.5 g of polymer in xylene at 135 °C (250 mL, stabilized with 1 mg 2,6-di-*tert*-butyl-4-methylphenol) and cooling to 25 °C. The precipitated solids were washed with xylene and isolated to yield the xylene-insoluble (XI) fraction.

^{13}C NMR (125.77 MHz) stereochemical analysis of the XI fractions was carried out using a Bruker DRX-500 spectrometer at 130 °C under proton noise decoupling in FT mode, with 10 KHz spectral width, 60 ° pulse angle and 15 s pulse repetition (2800 scans). A 5 mm water-cooled high temperature ^{1}H, ^{13}C dual probe suitable for measurements up to 600 °C was used. The sample concentration was 5-10 wt.-% in 1,2,4-trichlorobenzene containing 10 wt.-% tetradeutero-1,4-dichlorobenzene. Data reconciliation in the determination of the pentad distribution was carried out as previously described (10).

Results and Discussion

Effect of Hydrogen. The presence of hydrogen in propene polymerization usually leads not only to PP molecular weight reduction but also to significant increases in catalyst activity. This effect was first observed by Guastalla and Giannini (*11*) for the catalyst system $MgCl_2/TiCl_4$ - $AlEt_3$, while several groups (*12-21*) have reported on hydrogen activation effects observed using $MgCl_2/TiCl_4$/phthalate ester catalysts in combination with $AlEt_3$ as cocatalyst and an alkoxysilane as external donor.

Various proposals have been put forward to explain the observed increases in catalyst activity in the presence of hydrogen. These include oxidative addition of hydrogen to divalent titanium, regenerating active trivalent species (*22*), and reactivation of dormant π-allyl titanium species (*23*). An earlier claim (*13*) that the effect was due to an increase in the number of active species now appears to have been disproved, Bukatov et al. (*20*) having demonstrated that the presence of hydrogen actually decreased the active centre concentration but significantly increased the propagation rate constant. The dominant mechanism of hydrogen activation, applicable to both homogeneous and heterogeneous catalysts and concordant with an increase in the overall propagation rate constant, appears to be reactivation, via chain transfer, of "dormant" sites at which a regioirregular (2,1-) rather than the usual 1,2-insertion has taken place. Tsutsui et al. (*24*) have demonstrated this for a homogeneous metallocene catalyst, while Busico et al. (*25*) have shown that such sites were also formed in propene hydrooligomerization using a $MgCl_2$-supported catalyst.

The importance of secondary insertion in relation to chain transfer with hydrogen can be determined by ^{13}C NMR chain-end analysis of relatively low molecular weight polymers (*24*). Chain transfer after primary insertion gives an *i*-butyl chain-end, whereas a *n*-butyl group results from chain transfer after a secondary insertion, as illustrated below.

$$Ti\text{-}CH_2\text{-}CH(CH_3)\text{-}[CH_2\text{-}CH(CH_3)]_nPr + H_2 \longrightarrow Ti\text{-}H + i\text{-}Bu\text{-}CH(CH_3)[CH_2(CH_3)]_{n-1}Pr$$

$$Ti\text{-}CH(CH_3)\text{-}CH_2\text{-}[CH_2\text{-}CH(CH_3)]_nPr + H_2 \longrightarrow Ti\text{-}H + n\text{-}Bu\text{-}CH(CH_3)[CH_2(CH_3)]_{n-1}Pr$$

We have carried out chain-end analysis on the isotactic (xylene-insoluble, XI) fractions of polymers prepared at 67 - 70 °C in liquid monomer using $MgCl_2$-supported catalysts containing either diisobutyl phthalate (DIBP) or a diether as internal donor (*6,8*). Selected data are given in Table I, from which it can be seen that the proportion of *n*-Bu terminated chains is dependent on the catalyst system and on the amount of hydrogen in polymerization. The proportion of *n*-Bu terminated chains increases with decreasing (but finite) hydrogen pressure, which can be ascribed to a relatively high equilibrium concentration of "dormant" (2,1-inserted) species at low hydrogen pressure. It is important to note that these polymers have much lower molecular weights than most of the polypropylene grades which are produced commercially. This means that commercial polypropylene, produced at lower hydrogen pressures, will predominantly comprise polymer chains having a *n*-butyl terminal group, underlining the practical importance of chain transfer following the occasional secondary insertion.

Table I. **Effect of Electron Donors and Hydrogen on Polymer Molecular Weight, Stereoregularity and Chain-end Composition**

Internal Donor	External Donor	Hydrogen (bar)	[η] (dL/g)	*mmmm* in XI fraction (%)	*n*-Bu : *i*-Bu chain-ends in XI fraction
DIBP	CHMMS	2.5	1.41	97.5	46 : 54
"	"	8	1.05	98.0	24 : 76
DIBP	DCPMS	2.5	2.32	98.5	n.d.
"	"	8	1.35	99.0	34 : 66
diether	-	1	1.41	97.4	65 : 35
"	-	4	0.99	97.5	45 : 55

CHMMS: cyclohexylmethyldimethoxysilane
DCPMS: dicyclopentyldimethoxysilane
[η] :   intrinsic viscosity of total polymer

It is also apparent on comparing the effects of the external donors in Table I that dicylopentyldimethoxysilane (DCPMS) gives high molecular weight, highly stereoregular polymer and that for each catalyst system the polymer stereoregularity, evidenced by the isotactic (*mmmm*) pentad content, increases with increasing hydrogen concentration. This indicates that the probability of chain transfer is not only dependent on *regio*- but also on *stereo*selectivity, the presence of hydrogen leading in this case to an *i*-Bu terminated chain instead of a stereodefect in the polymer chain. We (*6,26*) and others (*20*) have suggested that this effect might be explained by a slowing down of chain propagation not only after a regio- but also after a stereoirregular insertion. However, as yet there is no evidence that a stereoirregular insertion does significantly slow down the propagation; an alternative explanation for the effect of hydrogen on chain stereoregularity could be that coordination of the "wrong" enantioface of the monomer may itself lead to a slowing down in chain propagation, leading to an increased probability of chain transfer.

A further important feature apparent from Table I is that, while the stereoselectivity of the diether-based catalyst used is seen to be similar to that of the DIBP/CHMMS system, the presence of the diether leads to significantly higher proportions of *n*-Bu terminated chains for polymers of similar molecular weight, indicating a high probability of chain transfer after secondary insertion in this system. Our studies have indicated (*8*) approximately one secondary insertion (followed by chain transfer with hydrogen) per 2000 monomer units in the isotactic polymer chain.

Effect of Monomer Concentration. Investigation of the effects of monomer concentration in propene polymerization can give important insight into the mechanism of selectivity control. For C_2-symmetric metallocene catalysts, polymer isotacticity decreases at low monomer concentrations due to a process of chain epimerization (*27,28*). With certain C_1-symmetric metallocenes however, for example isopropyl[(3-*t*-butylcyclopentadienyl)(fluorenyl)]zirconium dichloride (*29*), the situation is different. The presence of the bulky *t*-butyl substituent leads the growing chain to preferentially occupy the least hindered coordination position and to revert to this more enantioselective state via back-skip (*30*) after insertion. The probability of back-skip of the chain will increase with decreasing monomer concentration and in fact Rieger (*31*) has reported, for different C_1-symmetric metallocenes, an increase in stereoselectivity with decreasing propene concentration.

We have noted (*9*) that the stereosequence distribution of predominantly isotactic polypropylene prepared using C_1-symmetric metallocene catalysts deviates from that predicted by enantiomorphic-site statistics in that heptads such as *mrmrmm*, arising from isotactic sequences with two consecutive stereoerrors, are undetectable. For heterogeneous catalysts, the enantiomorphic-site model is only truly applicable if the active centre is so unsymmetric that systematic back-skip after insertion occurs, as indicated by Cossee (*32*), or if local C_2-symmetry is present (*33*). It now appears that the overall stereoselectivity of heterogeneous (MgCl$_2$-supported) catalysts does not correspond to either of these conditions, and that in this respect these catalysts exhibit behaviour similar to that of C_1-symmetric metallocenes.

This conclusion is based on both NMR spectroscopic evidence, indicating the absence of *mrmrmm* heptad sequences in the polymer (*9*), and on a determination of the effect of monomer concentration using various MgCl$_2$-supported catalysts. The results of polymerizations carried out at different monomer pressures, using the simple donor-free catalyst MgCl$_2$/TiCl$_4$ and a catalyst containing a diether as internal donor, are given in Table II. The reaction conditions (50 °C, no hydrogen) were chosen so as to obtain polymers having appreciable contents of stereochemical inversions, even with the diether-containing catalyst which in the presence of hydrogen and at higher temperature (*vide infra*) is highly stereoselective. The results show that, with both catalysts, a decrease in propene pressure leads to significant increases in stereoselectivity. This is apparent from higher proportions of isotactic (xylene-insoluble) polymer and increased isotacticity of the XI fractions at low propene pressure. Figure 1 illustrates the effect of monomer concentration on the *mmmm* pentad content of the XI fraction and of the overall polymer prepared using the relatively weakly isospecific catalyst system MgCl$_2$/TiCl$_4$ - AlEt$_3$, showing that even with this system fairly high isotacticity can be obtained at very low propene concentration.

Additional experiments were carried out with the catalyst system MgCl$_2$/TiCl$_4$/DIBP - AlEt$_3$ - alkoxysilane, as indicated in Table III. Various alkoxysilanes were used as external donors and in every case both the proportion of isotactic polymer and the stereoregularity of this fraction increased at low propene pressure.

Table II. Effect of propene pressure on catalyst stereo- and regioselectivity

Catalyst system	Propene pressure (bar)	Isotactic (XI) fraction			Xylene-soluble (XS) fraction		
		(wt.-%)	$mmmm$ (%)	2,1-units (%)	(wt.-%)	$mmmm$ (%)	2,1-units (%)
MgCl$_2$/TiCl$_4$ - AlEt$_3$	8	47.7	74.4	0.4	52.3	25.6	1.8
"	2.5	54.7	77.4	0.3	45.3	24.8	0.8
"	0.5	58.4	83.4	0.2	41.6	24.5	0.9
"	0.15	72.1	86.6	0.1	27.9	29.0	0.3
MgCl$_2$/TiCl$_4$/diether - AlEt$_3$	8	93.8	86.9	< 0.1	6.2	15.5	0.8
	0.2	96.8	93.9	< 0.1	3.2	38.4	0.1

Polymerization conditions: heptane slurry, 50 °C, no hydrogen

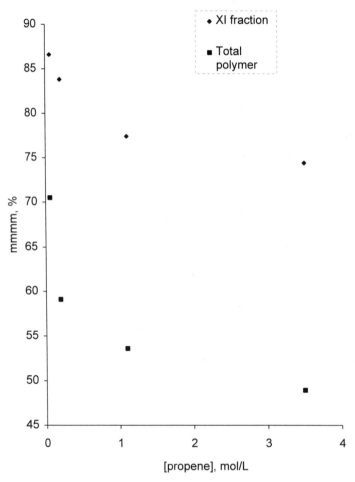

Figure 1. Effect of monomer concentration on content of *mmmm* pentads in polypropylene prepared with the catalyst system MgCl$_2$/TiCl$_4$ - AlEt$_3$. (Adapted from ref. 9.)

Table III. Effect of External Donor on the Relationship between Propene Pressure and Catalyst Stereoselectivity

External donor	Propene pressure (bar)	Isotactic (XI) fraction	
		(wt.-%)	*mmmm* (%)
none	8	76.4	81.9
"	0.2	96.4	90.6
PhSi(OEt)$_3$	8	87.8	87.9
"	0.2	94.2	94.7
CHMMS	8	90.1	89.6
"	0.2	93.2	94.8
DCPMS	8	92.8	90.4
"	0.2	95.7	95.1

Polymerization conditions: heptane slurry, 50 °C, no hydrogen
CHMMS: cyclohexylmethyldimethoxysilane
DCPMS: dicyclopentyldimethoxysilane

The above results indicate that an increase in stereoselectivity with decreasing monomer concentration is a general feature in propene polymerization with MgCl$_2$-supported catalysts. A similar effect with a MgCl$_2$/TiCl$_4$/diether - AlEt$_3$ catalyst system has previously been noted by Iiskola (*34*) and Paukkeri (*35*). It can be inferred that, in propene polymerization using heterogeneous Ziegler-Natta catalysts, at least part of the progagation occurs at unsymmetric active centres having C$_1$-type symmetry. The isotacticity of the polymer is therefore determined by the extent to which insertion at a highly enantioselective site is followed by chain back-skip as opposed to a less regio- and stereoselective insertion when the chain is in the coordination position previously occupied by the monomer.

Effect of Polymerization Temperature. It is well established that, in propene polymerization with MgCl$_2$-supported catalysts, the fraction of isotactic polymer increases with increasing temperature, at least for polymerization temperatures up to around 70 - 80 °C. On the other hand, surprisingly little has been reported on the effect of polymerization temperature on the stereoregularity of the isotactic fraction. Decreased stereoregularity with increasing temperature has been observed (*36*) using the catalyst system, MgCl$_2$/TiCl$_4$/dioctyl phthalate - AlEt$_3$ and was ascribed to progressive removal of the internal donor via interaction with the cocatalyst as the temperature was raised.

In order to determine the effect of polymerization temperature on the microtacticity of polypropylene prepared using $MgCl_2$-supported catalysts, we have carried out polymerizations with various catalyst systems in the range 20 - 80 °C, in the presence and absence of hydrogen (37). The results obtained using the catalyst system $MgCl_2/TiCl_4$/diether - $AlEt_3$ are given in Table IV. It is clear that increasing the polymerization temperature leads to significant increases in PP yield and in both the proportion and stereoregularity of the isotactic (xylene-insoluble) fraction. This was the case both in the presence and in the absence of hydrogen. As illustrated in Figure 2, the *mmmm* pentad content of the isotactic fraction increases with both increasing hydrogen concentration and polymerization temperature.

Table IV. Effect of temperature in propene polymerization using the catalyst system $MgCl_2/TiCl_4$/diether - $AlEt_3$

Temperature	Hydrogen	PP Yield	Isotactic (XI) fraction	
				mmmm
(°C)	(NL)	(kg/g cat.)	(wt.-%)	(%)
40	0	6	82.6	86.7
60	0	21	90.3	88.6
80	0	35	94.3	91.0
40	0.8	9	84.1	88.7
40	1.6	14	85.0	89.2
40	2.2	19	88.1	89.8
60	1.7	60	93.5	91.4
80	1.8	81	95.8	93.1
40	7.1	13	85.0	90.6
60	7.6	75	92.4	92.7
80	8.0	112	95.1	93.5

Polymerization conditions: liquid propene, 1 h.

Table V shows the effect of polymerization temperature on the microtacticity of polypropylene prepared using the catalyst system $MgCl_2/TiCl_4$/diisobutyl phthalate - $AlEt_3$ - external donor. These polymerizations were carried out in the absence of hydrogen, in order to avoid complications arising from the effect of hydrogen on polymer microtacticity. It is apparent that comparable increases in PP yield with increasing temperature are obtained with each system, while the XI values and the isotactic pentad contents of the XI fractions are dependent on both external donor and temperature. No increase in *mmmm* with increasing temperature is observed in the

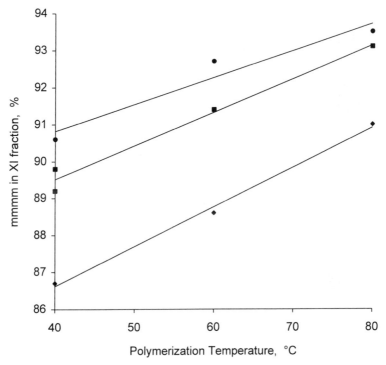

Figure 2. Effect of temperature on PP microtacticity, with the catalyst system
MgCl$_2$/TiCl$_4$/diether - AlEt$_3$.
(hydrogen added: ◆ 0; ■ 1.6-2.2; ● 7-8 NL)

Table V. Effect of Temperature in Propene Polymerization using the Catalyst System MgCl$_2$/TiCl$_4$/DIBP - AlEt$_3$ - External Donor

Temp. (°C)	External Donor	PP yield (kg/g cat.)	PP XI[a] (wt.-%)	[η][b] (dL/g)	Pentad distribution of xylene-insoluble fraction, (%)							
					mmmm	mmmr	rmmr	mmrr rmrr	rmrm	rrrr	mrrr	mrrm
20	-	1	67.9	3.61	80.3	4.6	0.7	2.4	0.4	2.5	2.1	2.1
40	-	3	73.3	3.62	81.7	4.5	0.7	1.9	0.5	1.8	1.4	2.4
60	-	8	66.0	2.93	82.6	4.8	0.6	1.6	0.4	1.4	1.2	2.4
80	-	10	64.4	2.05	81.2	4.9	0.7	1.8	0.7	1.2	1.2	2.7
20	TMPIP[c]	1	84.4	4.07	83.4	3.8	0.8	1.8	0.5	2.3	1.8	1.4
40	TMPIP	3	86.9	5.41	84.9	3.6	0.6	1.4	0.6	1.6	1.3	1.8
60	TMPIP	11	90.2	5.56	88.3	3.1	0.5	1.3	0.3	1.2	0.9	1.3
80	TMPIP	18	91.4	4.20	88.9	3.0	0.5	1.0	0.4	0.8	0.7	1.4
20	diether	1	85.9	3.46	86.2	3.1	0.7	1.5	0.4	1.5	1.3	1.6
40	diether	3	89.5	4.95	86.4	3.0	0.7	1.4	0.4	1.8	1.1	1.6
60	diether	6	95.0	5.70	91.0	2.0	0.6	0.9	0.3	1.0	0.6	1.1
80	diether	17	97.7	4.56	94.3	1.3	0.5	0.9	0.4	0.4	0.3	0.6
20	DCPMS[d]	1	88.0	4.94	87.8	2.9	0.5	1.4	0.4	1.2	1.2	1.4
40	DCPMS	5	95.0	6.49	90.8	2.2	0.4	1.0	0.3	1.0	0.7	1.1
60	DCPMS	7	95.6	6.40	91.4	2.2	0.5	0.9	0.3	0.6	0.5	1.1
80	DCPMS	14	97.1	6.61	94.9	1.3	0.2	0.6	0.1	0.3	0.3	0.8

a) PP XI: xylene-insoluble fraction of polypropylene.
b) Intrinsic viscosity
c) TMPIP: 2,2,6,6-tetramethylpiperidine.
d) DCPMS: dicyclopentyldimethoxysilane.

absence of external donor. This is in agreement with the results of Hsu (*36*). The pentad data for the polymers prepared in the absence of external donor reveal two contrasting trends as the polymerization temperature is increased. The proportion of isolated stereoinversions, evidenced by the pentads *mmmr*, *mmrr* and *mrrm*, increases with temperature and it is reasonable to ascribe this to loss of the (internal) donor as the temperature is raised. On the other hand, the proportion of the *rrrr* pentad decreases with increasing temperature. Syndiotactic sequences in PP prepared using MgCl$_2$-supported catalysts result from chain-end control (*38*). A decrease in *rrrr* pentad content may indicate shorter syndiotactic sequences, either as a result of a more rapid interconversion between isospecific and syndiospecific active species, or as a result of weaker chain-end control at elevated temperatures.

In the presence of external donor, the *mmmm* content of the XI fraction increases with increasing temperature. Figure 3 depicts the effect of temperature on the product of % XI and % *mmmm* in the XI fraction, which effectively represents the proportion of long isotactic sequences in the overall polymer. It is apparent that the steepest increases in microtacticity with increasing temperature are obtained with the diether-containing catalyst systems. Previous studies (*4,39,40*) have indicated that similar isospecific centres are present in these systems, irrespective of whether the ether is used as internal or external donor.

The above results demonstrate that, with the exception of the MgCl$_2$/TiCl$_4$/phthalate ester catalyst in the absence of external donor, an increase in polymerization temperature in polymerization leads to significant increases in the stereoregularity of the isotactic fraction. There is no indication that this effect is related to chain back-skip at a C$_1$-symmetric site, taking into account the fact that with highly isospecific C$_1$-symmetric metallocene catalysts only a (slight) *decrease* in isotacticity has been observed with increasing temperature (*41*). We ascribe the effect to the presence, in heterogeneous catalysts, of a range of active centres of different selectivity. The increase in microtacticity with increasing temperature can be explained by an increase in activation energy with increasing isospecificity of the active centre, leading to a proportionately greater increase in the productivity of highly isospecific as opposed to weakly isospecific sites as the polymerization temperature is raised.

This may be due to activation of strongly hindered sites, or to the effects of regioirregular insertion, if it can be assumed that the effect of a 2,1-insertion on slowing down chain propagation increases with site specificity. The relatively large effect of temperature on microtacticity observed with the diether-containing catalysts would be in line with a relatively high barrier to insertion following the occasional secondary insertion in these systems, as is indeed indicated by the high proportions of *n*-Bu terminated chains arising from chain transfer with hydrogen after secondary insertion in polymers prepared with diether-containing systems. It is noteworthy that microtacticity increases with temperature in the presence and in the absence of hydrogen. Therefore, if this explanation is correct it must be assumed that a 2,1-inserted centre remains in the dormant state for a significant period of time even in the presence of hydrogen, and that the use of a high temperature can effectively reduce the

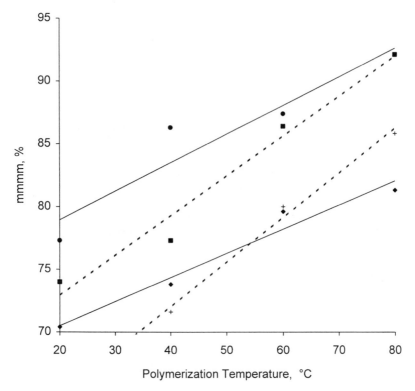

Figure 3. Effect of temperature on PP microtacticity, with the catalyst systems MgCl$_2$/TiCl$_4$/DIBP - AlEt$_3$ - external donor (◆ TMPIP; ■ diether; ● DCPDMS) and MgCl$_2$/TiCl$_4$/diether - AlEt$_3$ (+). The dotted lines indicate the trends obtained with the diether-containing systems.

barrier to chain propagation after 2,1-insertion. This implies that the molecular weight of the isotactic fraction may also increase with increasing polymerization temperature; investigation of this effect is currently in progress.

Conclusions

It is now evident that the performance of high-activity $MgCl_2$-supported catalysts used in polypropylene production, in terms of polymer yield, isotacticity, molecular weight and molecular weight distribution, is linked to the fundamental effects of stereo- and regioirregular insertion on chain propagation and chain transfer.

The probability of an irregular insertion is dependent on chain back-skip after insertion at unsymmetric active species having C_1-type symmetry and which comprise two non-equivalent coordination positions for the monomer and the growing chain, only one of which is highly enantioselective. The increase in isotacticity at low monomer concentration and the apparent absence of the *mrmrmm* heptad, which would arise from two consecutive stereoerrors in an isotactic sequence, indicates that occasional (isolated) misinsertions take place when chain migratory insertion at the enantioselective site is followed by an insertion at the less enantioselective site, rather than back-skip of the chain to the favoured position before the next insertion. Taking into account evidence that the (internal or external) electron donor is present in the environment of at least part of the active centres (*42,43*), it is clear that chain back-skip and therefore catalyst selectivity will be dependent on the (continued) presence of the donor. This in turn depends on the strength of coordination between the donor and the catalyst surface, for example a Mg atom adjacent to the active Ti. Polypropylene produced using heterogeneous Ziegler-Natta catalysts is essentially a stereoblock material, the isotacticity of a particular fraction depending on the relative block lengths of isotactic and weakly syndiotactic (syndiotactoid) sequences. This can be ascribed to rapid interconversion of the active species, particularly in the presence of donors which are weakly coordinating or have high molecular mobility.

Strong coordination and low molecular mobility is indicated in the case of the $MgCl_2/TiCl_4/DIBP$ - AlR_3 - alkoxysilane catalyst system where the external donor is dicyclopentyldimethoxysilane. This donor gives both high isotacticity and high molecular weight polymer. These features are connected, as the sensitivity of a Ziegler-Natta catalyst to chain transfer with hydrogen is strongly dependent on regio- and stereoselectivity. Chain transfer after regioirregular (2,1-) insertion is the most important mechanism determining the molecular weight of polypropylene produced commercially and is also the main factor responsible for the activating effect of hydrogen, due to the retarding effect of a secondary insertion on chain propagation.

Chain transfer with hydrogen after secondary insertion is particularly prominent with the newly-developed catalysts containing diether donors. In addition to having very high activity, these catalysts require relatively little hydrogen for molecular weight control. This effect appears to be due to a particularly large effect of an occasional 2,1-insertion on slowing down the propagation, the high "dormancy" of the resulting species increasing the probability of chain transfer.

Polymer microtacticity increases with increasing polymerization temperature, which can be ascribed to a greater relative increase in productivity for highly isospecific as opposed to moderately isospecific centres. This may be due to a lowering in the barrier to further propagation after secondary insertion at highly isospecific centres. Because of the fact that polymer molecular weight generally increases with site isospecificity, these considerations also provide an explanation for the fact that, with MgCl$_2$-supported catalysts, polymer molecular weight distribution narrows with increasing polymerization temperature. An increase in the proportion of polymer resulting from sites having the highest isospecificity implies that the fraction of relatively low molecular weight polymer produced on sites having lower specificity will decrease, leading to an overall narrowing in MWD.

In summary, it can be concluded that detailed polymer compositional and chain-end analysis has been instrumental in providing new insight into factors affecting the performance of high-activity MgCl$_2$-supported catalysts for polypropylene, particularly with regard to the profound effects of catalyst regio- and stereoselectivity.

References

1. Barbè, P.C.; Cecchin, G.; Noristi, L. *Adv. Polym. Sci.* **1987**, *81*, 1.
2. Parodi, S; Nocci, R.; Giannini, U.; Barbè, P.C; Scatà, U. *Eur. Pat.* 45977, **1982**.
3. Albizzati, E.; Barbè, P.C.; Noristi, L.; Scordamaglia, R.; Barino, L.; Giannini, U.; Morini, G. *Eur. Pat. Appl.* 361494, **1990**.
4. Albizzati, E.; Giannini, U.; Galimberti, M.; Barino, L.; Scordamaglia, R. *Macromol. Symp.* **1995**, *89*, 73.
5. Albizzati, E.; Giannini, U.; Morini, G.; Smith, C.A; Zeigler, R.C. In *Ziegler Catalysts. Recent Scientific Innovations and Technological Improvements*; Fink, G; Mülhaupt, R.; Brintzinger, H.H., Eds.; Springer-Verlag: Berlin, **1995**, pp 413-425.
6. Chadwick, J.C.; van Kessel, G.M.M.; Sudmeijer, O. *Macromol. Chem. Phys.* **1995**, *196*, 1431.
7. Collina, G; Morini, G.; Ferrara, G. *Polym. Bull. (Berlin)* **1995**, *35*, 115.
8. Chadwick, J.C.; Morini, G.; Albizzati, E.; Balbontin, G.; Mingozzi, I.; Cristofori, A.; Sudmeijer, O. *Macromol. Chem. Phys.* **1996**, *197*, 2501.
9. Busico, V.; Cipullo, R.; Talarico, G.; Segre, A.L.; Chadwick, J.C. *Macromolecules* **1997**, *30*, 4786.
10. Van der Burg, M.W.; Chadwick, J.C.; Sudmeijer, O.; Tulleken, H.J.A.F. *Makromol. Chem., Theory Simul.* **1993**, *2*, 385.
11. Guastalla, G.; Giannini, U.; *Makromol. Chem., Rapid Commun.* **1983**, *4*, 519.
12. Spitz, R.; Bobichon, C.; Guyot, A. *Makromol. Chem.* **1989**, *190*, 717.
13. Parsons, I.W.; Al-Turki, T.M. *Polym. Commun.* **1989**, *30*, 72.
14. Kioka, M.; Kashiwa, N.; *J. Macromol. Sci., Chem.* **1991**, *A28*, 865.
15. Albizzati, E., Galimberti, M.; Giannini, U.; Morini, G. *Makromol. Chem., Macromol. Symp.* **1991**, *48/49*, 223.
16. Imaoka, K.; Ikai, S.; Tamura, M.; Yoshikiyo, M; Yano, T. *J. Molec. Catal.* **1993**, *82*, 37.

17. Chadwick, J.C.; Miedema, A.; Sudmeijer, O. *Macromol. Chem. Phys.* **1994**, *195*, 167.
18. Bukatov, G.D.; Goncharov, V.S.; Zakharov, V.A.; Dudchenko, V.K.; Sergeev, S.A. *Kinet. Catal.* **1994**, *35*, 358.
19 Rishina, L.A.; Vizen, E.I.; Sosnovskaja, L.N.; Dyachkovsky, F.S. *Eur. Polym. J.* **1994**, *30*, 1309.
20. Bukatov, G.D.; Goncharov, V.S.; Zakharov, V.A. *Macromol. Chem. Phys.* **1995**, *196*, 1751.
21. Kojoh, S.; Kioka, M.; Kashiwa, N.; Itoh, M.; Mizuno, A. *Polymer* **1995**, *36*, 5015.
22. Chien, J.C.W.; Nozaki, T. *J. Polym. Sci., Part A: Polym. Chem.* **1991**, *29*, 505.
23. Guyot, A.; Spitz, R.; Dassaud, J.-P.; Gomez, C. *J. Molec. Catal.* **1993**, *82*, 29.
24. Tsutsui, T.; Kashiwa, N.; Mizuno, A. *Makromol. Chem., Rapid Commun.* **1990**, *11*, 565.
25. Busico. V.; Cipullo, R.; Corradini, P., *Makromol. Chem., Rapid Commun.* **1992**, *13*, 15.
26. Busico, V.; Cipullo, R.; Chadwick, J.C.; Modder, J.F.; Sudmeijer, O. *Macromolecules* **1994**, *27*, 7538.
27. Busico, V.; Cipullo, R. *J. Amer. Chem. Soc.* **1994**, *116*, 9329.
28. Resconi, L.; Fait, A.; Piemontesi, F.; Colonnesi, M.; Rychlicki, H.; Zeigler, R. *Macromolecules* **1995**, *28*, 6667.
29. Ewen, J; Elder, M.J. *Eur. Pat. Appl.* 537130, **1993**.
30. Guerra, G.; Cavallo, L.; Moscardi, G.; Vacatello, M.; Corradini, P. *Macromolecules* **1996**, *29*, 4834.
31. Rieger, B.; Jany, G.; Fawzi, R.; Steinmann, M. *Organometallics* **1994**, *13*, 647.
32. Cossee, P. In *The stereochemistry of macromolecules*; Ketley, A.D., Ed.; Marcel Dekker: N.Y., **1967**, Vol. 1; pp 145-175.
33. Corradini, P.; Busico, V.; Cavallo, L.; Guerra, G.; Vacatello, M.; Venditto, V. *J. Molec. Catal.* **1992**, *74*, 433.
34. Iiskola, E.; Pelkonen, A.; Kekkonen, H.J.; Pursiainen, J.; Pakkanen, T.A. *Makromol. Chem., Rapid Commun.* **1993**, *14*, 133.
35. Paukkeri, R.; Iiskola, E.; Lehtinen, A.; Salminen, H. *Polymer* **1994**, *35*, 2636.
36. Yang, C.B.; Hsu, C.C. *J. Appl. Polym. Sci.* **1995**, *58*, 1237.
37. Chadwick, J.C.; Morini, G.; Balbontin, G.; Sudmeijer, O. *Macromol. Chem Phys.* **1998**, in press.
38. Busico, V.; Corradini, P.; De Martino, L.; Graziano, F.; Iadicicco, A. *Makromol. Chem. Phys.* **1991**, *192*, 49.
39. Sacchi, M.C.; Forlini, F.; Tritto, I,; Locatelli, P.; Morini, G.; Noristi, L.; Albizzati, E. *Macromolecules* **1996**, *29*, 3341.
40. Chadwick, J.C.; Morini, G.; Balbontin, G.; Mingozzi, I.; Albizzati. E.; Sudmeijer, O. *Macromol. Chem. Phys.* **1997**, *198*, 1181.
41. Miyake, S.; Okumura, Y.; Inazawa, S. *Macromolecules* **1995**, *28*, 3074.
42. Sacchi, M.C.; Forlini, F.; Tritto, I.; Locatelli, P.; Morini, G.; Baruzzi, G.; Albizzati, E. *Macromol. Symp.* **1995**, *89*, 91.
43. Morini, G.; Albizzati, E.; Balbontin, G.; Mingozzi, I.; Sacchi, M.C.; Forlini, F.; Tritto, I. *Macromolecules* **1996**, *29*, 5770.

Chapter 5

Homo- and Copolymerization of Styrene by Bridged Zirconocene Complex with Benz Indenyl Ligand

Toru Arai, Shigeru Suzuki, and Toshiaki Ohtsu

Research Center, Denki Kagaku Kogyo Company, Ltd.,
3–5–1 Asahimachi, Machida-city, Tokyo 194–8560, Japan

Rac-[isopropylidenebis(4,5-benzindenyl)] zirconium dichloride/ MAO promotes copolymerization of ethylene and styrene to form a random copolymer with a high activity. The random copolymer has head to tail St-St sequences and isotactic stereo-regularity both of the head to tail St-St and alternating Et-St sequences. This catalytic system also promotes homo-polymerization of styrene to form isotactic polystyrene. The steric interaction of the C_2 symmetric ligand framework to polymer chain ends and monomer phenyl groups may control isotactic propagation, similarly to the isotactic polymerization of propene reported, with complexes having C_2 symmetry.

Stereoregular polymerization of styrene to form isotactic and syndiotactic polystyrene has been reported using the conventional Zieglar-Natta catalyst and half metallocene or non-metallocene type catalyst systems combined with methylalumoxane (MAO) (1,2), respectively. However, there have been no reports of stereo-regular homo-polymerization of styrene using bridged metallocene type single site catalysts, which are known to promote stereo-regular polymerization of α-olefins.

Recently, single-site catalyzed ethylene-styrene (Et-St) copolymerization has been reported. [Dimethylsilandyl(tetramethylcycropentadienyl)(t-butylamido)] titanium dichloride (CGCT-type catalyst) (3), [isopropylidene(9-fluorenyl)(1-cyclopentadienyl)] zirconium dichloride (Ewen-type zirconocene complex) (4) and CpTiCl$_3$ (5) activated by methylalumoxane (MAO) produce "pseudo-random" Et-St copolymers (3). These copolymers have no head to tail St-St sequences, which limit the St content below 50 mol%, and no stereo-regularity of the phenyl group in the alternating Et-St structure. More recently, Oliva et al. have reported a nearly alternating Et-St copolymer with stereo-regularity under low temperature conditions (-25℃) using rac-[ethylenebis(indenyl)]zirconium dichloride (Brintzinger-type zirconocene complex) (*6*). The copolymer shows no head to tail sequences in the 13C-NMR spectrum and, in our investigation, the complex could not produce a high St copolymer having more than about 10 mol % at 50℃.

On the other hand, restricted "alternating" Et-St copolymer can be produced by Ti-complexes with thio-bis phenol ligand (*7*). This copolymer consists only of an isotactic alternating Et-St structure, so that the St content is limited to about 50mol%. This explains why there are no reports of St-Et copolymer containing styrene more than 50 mol%, due to the difficulty of the head to tail St-St sequences formation.

This report presents complete "random" St-Et copolymerization with the head to tail St-St sequences and stereo-regular sequences, and styrene homo-polymerization to form isotactic polystyrene by using an isopropylidene-bridged benzindenyl zirconocene catalyst. These results suggest a new stereo-regulation mechanism for styrene polymerization using a single-site catalyst.

Experimental

Complexes   Complexes used are shown in figure 1. [Dimethylsilandyl(tetramethyl cycropentadienyl)(t-butylamido)]titanium dichloride (Ⅱ :CGCT-type) (*3*), [diphenylmethylene(fluorenyl)(cycropentadienyl)]zirconium dichloride (Ⅲ :Ewen-type) (*8*), and rac-[ethylenebis(indenyl)]zirconium dichloride (Ⅳ :Brintzinger-type) (*9*) were produced by the methods reported in the literature.

Rac-[isopropylidenebis(4,5-benzindenyl)]zirconium dichloride Ⅰ was prepared as follows. 4,5-Benzindene {benz(e)indene} was prepared according to a published method (*10*). 1,1-isopropylidene-4,5-benzindene was produced form 4,5-benzindene and acetone (*11*). The ligand system of isopropylidene bis-benzindene was synthesized using the reaction of Li salt of 4,5-benzindene and 1,1-isopropylidene-4,5-benzindene in THF at R.T. The reaction of the ligand and Zr(NMe₂)₄ in Toluene at 130℃ for 10 hr resulted in the formation of a di-amide complex. Crude complex was obtained by removing the solvent, after the reaction of di-amide complex and Me₂NH-HCl in CH₂Cl₂ at -78℃, and purified by washing with pentane and a small amount of THF to give a yellow micro-crystalline form of the complex Ⅰ (racemic-form).

1H-NMR.   (CDCl$_3$, 25℃) δ =8.01(m, 2H), 7.75(m, 2H), 7.69(d, 2H), 7.48-7.58(m, 4H), 7.38(d, 2H), 7.19(d, 2H), 6.26(d, 2H), 2.42(s, 6H).

Mass-Spectrometry (MS).   The accurate molecular weights of the products were confirmed by time-of-flight (TOF) mass-spectrometry, using a Micromass TofSpec E MALDI-TOF system. The instrument was operated at an accelerating voltage of 20-25 kV in the reflectron mode, using a time-lag focussing source and N$_2$ UV laser (λ =337nm, 175 μ J). Without employing a matrix, the complex gave strong, isotopically resolved ions at average m/z=532 and m/z=495, allowing confirmation of the composition by the isotopic distributions (C$_{29}$H$_{22}$ZrCl$_2^+$ or C$_{29}$H$_{22}$ZrCl$^+$, respectively).

Anal.   Calcd for C$_{29}$H$_{22}$ZrCl$_2$, C, 65.39% ; H, 4.16%.   Found: C, 63.86% ; H, 3.98%.

68

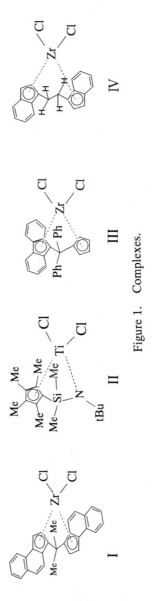

Figure 1. Complexes.

Copolymerization. Polymerization runs were conducted in a 10L size autoclave equipped with a mechanical stirrer and heating-cooling jacket. Styrene homo-polymerization was carried out using 120 ml size glass tube. MAO (Methylalumoxane M- or P-MAO) and tri-isobutylaluminium were purchased from Tosoh-akzo and Kanto-Chemicals, respectively.

Characterization. The 13C-NMR spectra of the polymers were measured using an alfa-500 spectrometer with $CDCl_3$ or 1,1,2,2-tetrachloloethane with TMS as the standard. The signals were assigned according to references (5,7,12), DEPT-method, improved Grant and Paul empirical method (13), and database 'SPECINFO' (14). Molecular weights were obtained by GPC (HLC-8020) equipped with RI and UV detectors. St contents of the copolymers were determined from 1H-NMR spectra. The DSC measurements were carried out using a SEIKO DENSHI DSC-200 under N_2 flowing at a temperature rate of 10℃/min.

Results and Discussion

Table-1 shows the typical copolymerization results. Complex I shows a markedly high St-Et copolymerization activity as compared with the II(CGCT-), III(Ewen-), and IV(Brintzinger-type) complexes under the same conditions. The single and narrow peaks (molecular weight distributions about 2.0) obtained by GPC measurement indicates the single-site catalysis. The peaks with the same pattern measured by RI and UV detectors also show a high uniformity of the St distribution in the copolymers (figure 2). This complex could form copolymers with styrene contents from 10 to more than 70 mol % by changing the monomer feed ratio.

The 13C-NMR spectra of the copolymer (run 1, 2, 4 and 5) are shown in fig.3a, b, c, and d. Peaks at 25.1, 36.5, and 45.3 ppm attributed to the carbon atoms in the alternating St-Et sequences and 27.5, 29.6, 36.7, and 45.6-8 ppm from Et-Et sequences with a styrene unit are observed. New peaks at 42.9 and 43.9 ppm are assigned to the head to tail sequences of the 2 styrene units, and 40.9 and 43.5 ppm are from isotactic polystyrene sequences consisting of more than 3 styrene units (Scheme I). However, there are no peaks in the same region of the "pseudo random" copolymers obtained by II, III, and IV, as shown in fig. 4 and as already reported (3,4,6).

New peaks in the vicinity of 24.8 and 36.0 ppm obtained by I (run 4 and 5) were attributed to the methylene carbon by the DEPT method of 13C-NMR spectroscopy. The intensities increased with incerase in St content and the intensities of the peaks attributed to the earlier-mentioned head to tail St-St sequences. From these results, we assigned the new speaks to the structure of the isolated ethylene unit by styrene sequences (Scheme I). The shift in peak position from 25.1 ppm (S β β carbon in alternating structure) to 24.8 ppm (S β β + carbon in isolated ethylene unit by styrene sequence) may be due to the phenyl group on the δ carbon.

We thus concluded that it is a "random copolymer", since it has the head to tail St-St sequences, besides the Et-Et and St-Et sequences. The formation of higher St content copolymer than 50 mol % is due to the head to tail St-St sequence formation ability of the complex I .

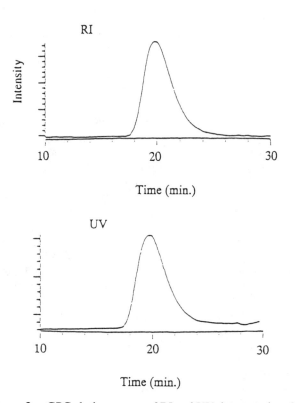

Figure 2.   GPC elution curves of RI and UV detectors (run 3).

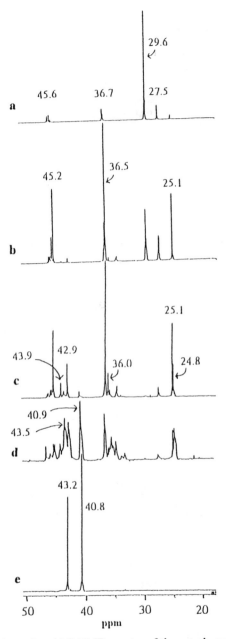

Figure 3. 13C-NMR spectra of the copolymers.
(TMS scale, C₂D₂Cl₄, Methine, methylene region)
a: run 1 (St 11.5 mol%), b: run 2 (St 37.1 mol%),
c: run 4 (St 55.5 mol%), d: run 5 (St 72.9 mol%),
e: run 6 (St 100 mol%).

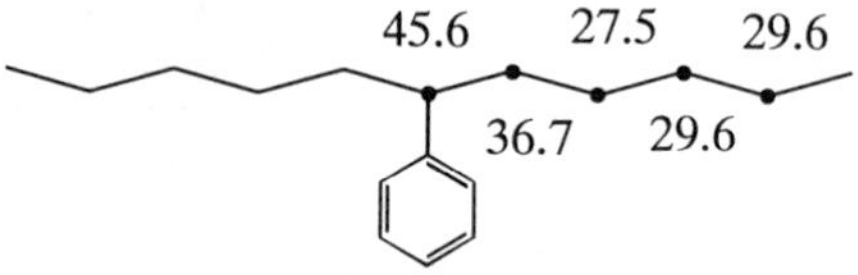

Et-Et sequences with a St unit.

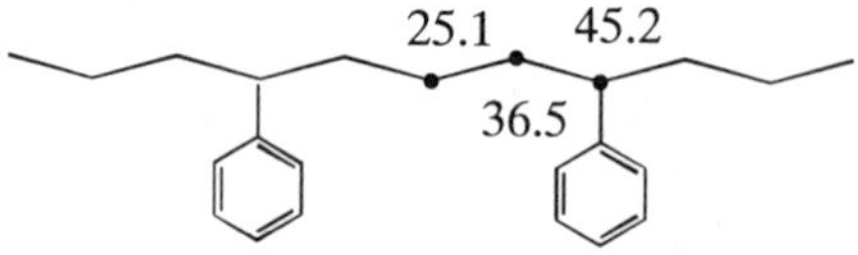

Alternating St-Et sequence. (Isotactic)

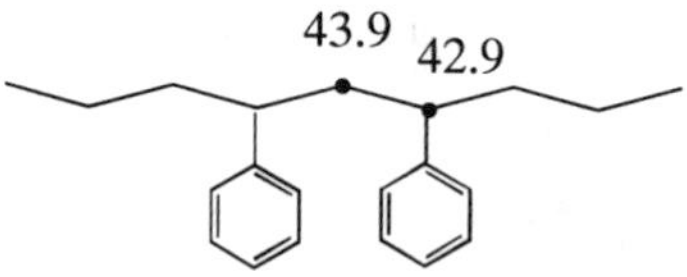

Head to tail St-St sequence (2 St units).

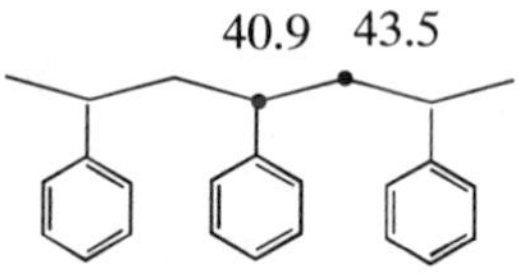

Isotactic polystyrene sequences.
(More than 3 St units)

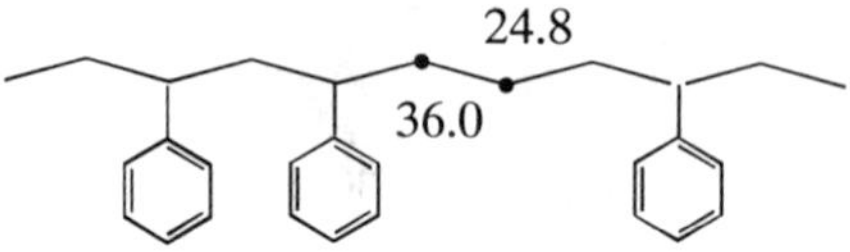

Isolated Ethylene unit by St sequences.

Scheme I   Main sequences in copolymer produced by complex I.
(Peak sift values of 13C-NMR)

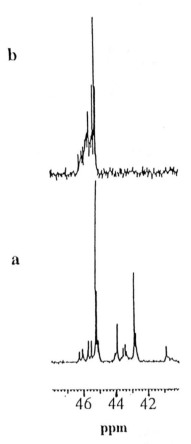

Figure 4.   13C-NMR spectra of the copolymers. (40 - 46ppm)
a: run 4 (complex Ⅰ), b: run 8 (complex Ⅱ).

S β β carbon of the St-Et alternating structure produced by complex I shows only a peak assigned to a meso-diad structure (25.1 ppm), indicating a highly isotactic alternating structure as shown in fig.5, which is similar to the copolymer produced by rac-[isopropylidenebis(indenyl)]zirconium dichloride, as already reported (15). The copolymer produced by complex IV, in spite of the low styrene content and molecular weight, has an isotactic alternating structure (table-1). In contrast, complexes II and III(4) give an atactic structure having both meso- and rac-diad peaks. From these results, a C_2 symmetric framework with indenyl or substituted indenyl groups seems to be a key structure for stereo-regulation under the polymerization conditions.

Complex I with MAO performs homo-polymerization of styrene (run 6) to form isotactic polystyrene, as shown in fig.3e (13-C NMR spectrum) and fig.6 a (XRD spectrum). To our knowledge, this is the first observation of the homo-polymerization of styrene with isotactic stereo-regularity by the catalytic system of metallocene - MAO.

Copolymers with St content from 10 to 55 mol% show melting transitions of about 100°C in accordance with DSC measurement. The same X-ray diffraction patterns with the restricted alternating copolymer reported by Kakugo et al (7)., were obtained after the annealing procedure, with the St content being about 20-55 mol% (figure 6, b and c). This indicates that the isotactic alternating sequence in the random structure will be responsible for the crystallinity. Relatively low melting transitions can be accounted for by the existence of other sequences such as Et-Et or St-St. However, no diffraction peak was observed in the copolymer produced by complex II (figure 6 d). Single glass transitions increase linearly with St content, as shown in fig.7, which corresponds to the high uniformity of the copolymer produced by I.

A stereo-control mechanism for syndio-specific propagation of styrene with CpTiCl$_3$/MAO catalyst system has been proposed (16). The interaction between coordinated phenyl groups of the polymer chain end and styrene monomer, and cyclopentadienyl ligand to the titanium cation determines the stereochemistry. However, the 14 electron cation species of complex I, formed by contact with MAO, prevents the coordination of the phenyl groups to the Zr cation. This might suggest that the steric interaction of the C_2 symmetric ligand framework on polymer chain end and monomer phenyl groups control the isotactic propagation, similar to the isotactic polymerization of propene reported with complexes having C_2 symmetry (9,17).

In the case of the C_2 symmetric complexes, less or no repulsive interaction between the polymer chain end and 2,1-coordinated monomer phenyl groups (18) may lead to the formation of the head to tail St-St sequences (Scheme II). In contrast, Cs symmetric frameworks such as II and III may result in a repulsive interaction, inhibiting the formation of the head to tail St-St sequences (19).

The high isotacticity of the alternating Et-St sequence could be accounted for by the same mechanism as that for the isotactic St-St sequence formation, as described in Scheme III (19). C_2 symmetry of complex seems to favor the isotactic propagation of the St-Et alternating sequence, since the tacticity is not influenced by

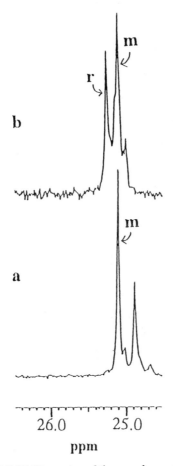

Figure 5.   13C-NMR spectra of the copolymers. (S β β region)
a: run 4 (complex I), b: run 8 (complex II).

Table1. Polymerization conditions and results.

Run No.	Complexes (μmol)	MAO (mmol)	St / Tol (ml / ml)	Et (MPa)	Time (h)	Yield (g)	Prod. a) (t/mol)	Mw (/10^4)	Mw/Mn	St Cont. (mol%)	Tacticity c)	Tm. e) (°C)
1	I ; 0.84	8.4	800 / 4000	1.1	5.0	464	550	18.5	2.2	11.5	m > 0.95	75
2	I ; 8.4	84	2400 / 2400	1.1	1.5	1320	157	33.0	2.3	37.1	m > 0.95	103
3	I ; 21	84	4000 / 800	0.2	6.0	1000	48	24.6	2.0	53.8	m > 0.95	102
4	I ; 21	84	4000 / 800	0.15	8.0	1010	48	16.9	1.9	55.5	m > 0.95	110
5	I ; 84	84	4000 / 800	0.02	5.0	150	1.8	5.2	1.9	72.9	m > 0.95	72
6	I ; 1.5	8.4	10 / 10	0	3.0	1.5	1.0	3.6	1.9	100	mmmm >0.9 d)	222
7	II ; 21	84	1500 / 3300	1.1	2.5	550	26.2	19.0	1.6	13.0	m = 0.5	63
8	II ; 84	84	4000 / 800	0.2	3.0	570	6.8	35.3	2.1	49.8	m = 0.5	n.d.
9	III ;164	164	4000 / 800	0.4	4.0	286	1.7	50.2	2.7	21.1	m = 0.65	25
10	IV ; 84	84	4000 / 800	0.2	6.0	386	4.6	5.0	2.0	9.0	m ≒ 0.9	112

Polymerization 50℃, except for run 6 (23 ℃). 8.4 mmol of TIBA (triisobutylaluminum) was charged, except for run 6 (1.0 mmol).

a) Productivity (ton-polymer/mol-complex) b) By 1H-NMR

c) Meso-diad of St-Et alternating structure (S β β carbon) d) Meso-pentad of polystyrene (Ph C1 carbon) e) By DSC.

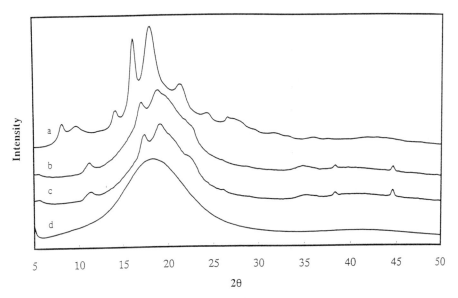

Figure 6.   X-ray diffraction spectrum of the copolymers. (Cu K α)
a: run 6, b: run 2, c: run 3, d: run8.

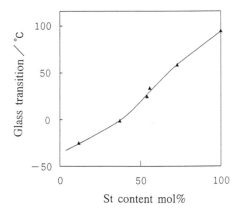

Figure 7.   Relation between St contents and glass transitions.

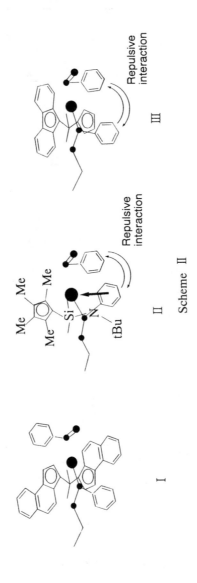

Scheme II

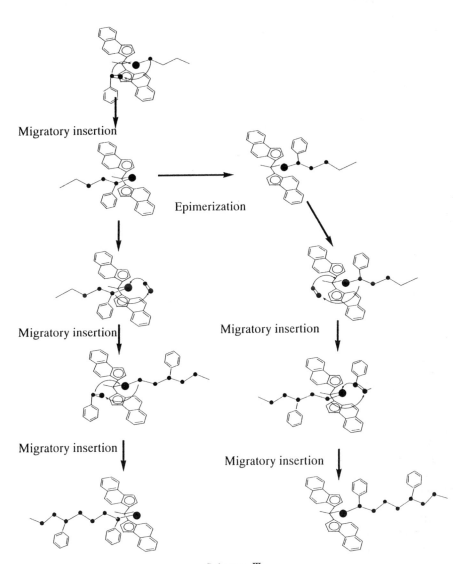

Migratory insertion

Epimerization

Migratory insertion

Migratory insertion

Migratory insertion

Migratory insertion

Scheme Ⅲ

migration errors such as site epimerization. Recently, Oliva et al. (*20*) reported a nearly isotactic alternating copolymer using [isopropylidene(9-fluorenyl)(1-cyclopentadienyl)] zirconium dichloride (Ewen-type) - MAO catalytic system at 0°C. However, the reference's (*4*) result shows an atactic alternating sequence formation under 40°C. The higher temperatures may allow or enhance the site migration error (site epimerization), which would cause atactic alternating sequence formation by using Cs symmetric complex.

References

1. Ishihara, N.; Kuramoto, M.; and Uoi, M. *Macromolcules,* **1988**, 21, 3356.
2. Pellecchia, C.; Longo, P.; Grassi, A.; Ammendola, P.; Zambelli, A. *Makromol. Chem., Rapid Commun.,* **1987**, 8, 277.
3. a) Stevens, J. C.; Timmers, F. J.; Wilson, D. R.; Schmidt, G. F.; Nickias, P. N.; Rosen, R. K.; Knight, G. W.; Lai, S. -y. (The Dow Chemical Company) , Eur. Pat. Appl. 0416815A2 (1990). b) Sernetz, F. G.; Mulhaupt, R.; Amor, F.; Eberle, T.; Okuda, J. *J. Polym. Sci., Part A : Polym. Chem.,* **1997**, 35, 1571.
4. Inoue, N.; Shiomura, T. *Polymer Preprints, Japan,* **1993**, 42, 2292.
5. Longo, P.; Grassi, A.; Oliva, L. *Makromol. Chem.,* **1990**, 191, 2387.
6. Oliva, L.; Izzo, L.; Longo, P. *Macromol. Rapid Commun.,* **1996**, 17, 745.
7. Kakugo, M.; Miyatake, T.; Mizunuma, K. *Stud. Surf. Sci. Catal.,* **1990**, 517.
8. Razavi, A.; Atwood, J. L. *J. Organomet. Chem.,* **1993**, 459, 117.
9. Kaminsky, W.; Kulper, K.; Brintzinger, H. H.; Wild, F. W. W. P. *Angew. Chem., Int. Ed. Engl.,* **1985**, 24, 507.
10. Stehling, U.; Diebold, J.; Kirsten, R.; Röll, W.; Brintzinger, H. H.; Jüngling, S.; Mülhaupt, R.; Langhauser, F. *Organometallics,* **1994**, 13, 964.11.
11. Yates, P.; Kronis, J. D. *Can. J. Chem.,* **1984**, 62, 1751.
12. Suzuki, T.; Tsuji, Y.; Watanabe, Y.; Takegami, Y. *Macromolecules,* **1980**, 13,849.
13. Randall, J. C. *J. Polymer Sci., Polymer Phys. Ed.,* **1975**, 13, 901.
14. The peak shift prediction by the 13C-NMR data base STN (SPECINFO).
15. Arai, T.; Ohtsu, T.; Suzuki, S. *Polymer Preprints,* **1997**, 38(2), 349. *Macromol. Rapid commun.,* 1998, 19, 327.
16. Longo,P.; Proto, A.; Zambelli, A. *Macromol. Chem.Phys.,* **1995**, 196, 3015.
17. Ewen, J. A. *J. Am. Chem. Soc.,* **1984**, 106, 6355. Ewen, J. A.; Elder, M. J.; Jones, R. L.; Haspeslagh, L.; Atwood, J. L.; Bott, S. G.; Robinson, K. *Makromol. Chem., Macromol. Symp.,* **1991**, 48/49, 253.
18. Oliva, L.; Caporaso, L.; Pellecchia, C.; Zambelli, A. *Macromolecules,* **1995**, 28, 4665.
19. Arai, T.; Suzuki, S.; Ohtsu, T. *submitted.*
20. Oliva, L.; Longo, P.; Izzo, L. *Macromolecules,* **1997**, 30, 5616.

Chapter 6

Random Copolymerization of Propylene and Styrene with Homogeneous Monocyclopentadienyltitanium–Methylaluminoxanes Catalyst

Qing Wu, Qinghai Gao, Zhong Ye, and Shangan Lin

Institute of Polymer Science, Zhongshan University, Guangzhou 510275, China

Copolymerizations of propylene and styrene in the presence of cyclopentadienyltitanium tribenzyloxide (CpTi(OBz)$_3$) and different methylaluminoxanes (MAO) have been investigated. It was found that the composition and structure of the polymerization products are strongly dependent on the amount of free alkylaluminium, involving the residual trimethylaluminum (TMA) in MAO and external one. Propylene-styrene copolymer could be prepared only with the catalyst system containing less TMA. For the copolymerization catalyzed with the titanocene and MAO containing the residual TMA 12.8 mol%, monomer reactivity ratios were estimated to be $r_P = 4.6$ and $r_S = 0.11$. The copolymers contain certain amount of regioirregular monomer-alternating sequences.

Since Sinn and Kaminsky (*1*) reported the direct synthesis of methylaluminoxane (MAO) by controlled hydrolysis of trimethylaluminum (TMA), a variety of metallocene catalyst systems for α-olefin polymerizations has been reported (*2-7*). Studies on catalysts applying in polymerizations of ethylene and propylene have so far focused mainly on the metallocenes with binary η^5-ligands (identical or not identical). On the other hand, the half-metallocenes, especially half-titanocenes, have been recognized to be efficient catalyst precursors for the syndiospecific polymerization of styrene as well as substituted styrene (*6*), conjugated diene (*8-9*), and ethylene (*3*) and their copolymerization each other (*6, 10-14*). However, very little work on propylene polymerization (*15*) and copolymerization of propylene and styrene with the monocyclopentadienyl catalysts has been reported in open literature.

Copolymerization of propylene and styrene with heterogeneous Ziegler-Natta catalysts was reported by Soga *et al.* (*16*). The catalyst system composed of Solvay-type $TiCl_3$ with Cp_2TiMe_2 as cocatalyst afforded isotactic PP, and afforded random copolymers only with styrene unit content less than 5% when the monomer mixture was copolymerized. However, an attempt to produce propylene-styrene copolymer and other α-olefin-styrene copolymers with homogeneous $Ti(OMen)_4$-MAO catalyst system failed (*17*).

It is well known that MAO prepared by controlled hydrolysis of TMA always contains some amount of unreacted TMA that can not be removed easily by vacuum distillation. Recently, we have performed styrene polymerization (*18* and Wu, Q. *J. Appl. Polym. Sci.*, in press) and propylene polymerization (19) catalyzed by mono(η^5-cyclopentadienyl)titanium complex combined with MAO as cocatalyst, and found that the catalytic activity depends upon the content of the residual trimethylaluminium (TMA) in MAO to a great extent.

In the present work, we have made a comparison between propylene and styrene polymerizations with the $CpTi(OBz)_3$-MAO catalyst system, and tried to apply the half-titanocene/MAO catalyst system in the synthesis of propylene-styrene copolymer.

Experimental Methods

Materials. All work involving air- and moisture-sensitive compounds was carried out under a dry nitrogen atmosphere. Toluene was dried by refluxing over sodium-benzophenone and distilled just before use. Styrene was distilled from calcium hydride under reduced pressure and stored in the darkness under nitrogen. Polymerization grade propylene was further purified before feeding to the reactor by conducting the gaseous monomer through columns containing molecular sieves and MnO. TMA was commercially available and used without further purification.

$CpTi(OBz)_3$ was synthesized as described previously (*18*). MAOs were prepared as follows: 200 mL of TMA solution (3.1M in toluene) was added dropwise into a flask with appropriate amount of ground $Al_2(SO_4)_3.18H_2O$ in toluene at 0 °C. The mixture was gradually heated to 60 °C and stirred for 24 h, and was then filtered. The filtrate was concentrated under reduced pressure to a white solid. Residual TMA content in the MAO was determined by pyridine titration. The MAOs prepared with initial $[H_2O]/[TMA]$ molar ratios of 1.8 and 1.0 contain the residual TMA 12.8% and 28.0%, respectively, and are referred to MAO1 and MAO2.

Polymerization Procedure. Copolymerizations were carried out in a 100 mL glass flask equipped with a magnetic stirrer. In the reactor filled with propylene, 20 mL of toluene and desired amounts of styrene, MAO and $CpTi(OBz)_3$ were introduced in this order. Gaseous propylene was continuously fed to replenish the consumed propylene monomer and maintain a constant pressure throughout the course of the copolymerization. The polymerizations were stopped after 1 h and the polymers were precipitated by addition of acidified alcohol. The resulting polymers were washed with alcohol and dried in vacuo to constant weight.

Analysis and Characterization. Oxidation states of titanium in catalyst systems were measured by redox titration (20). Extraction of the crude polymers with heptane was made in a Soxhlet extractor. ^{13}C NMR spectra of the polymers in chloroform were recorded on a Bruker-AM400 spectrometer operating at 50 °C under proton decoupling in the Fourier transform mode. The chemical shifts are reported in ppm downfield from TMS. Differential scanning calorimetric (DSC) curves of the polymers were obtained on a Perkin-Elmer DSC-7c instrument at a scanning rate of 10 °C/min from -50 °C to 300 °C.

Propylene and Styrene Homopolymerizations

For a comparison, homopolymerizations of propylene and styrene were carried out in the presence of CpTi(OBz)$_3$ and MAO担 containing different amounts of residual TMA. MAO1 contains residual TMA 12.8% and MAO2 contains 28%. Table I shows the results from the homopolymerizations. For styrene polymerization, the catalyst system with MAO2 exhibited an excellent activity about 11 times higher than that catalyst system with MAO1. On the contrary, for propylene polymerization with MAO2 only trace polymer was obtained.

Table I. Homopolymerizations of Propylene and Styrene with CpTi(OBz)$_3$-MAO Catalyst Systems

MAO	Residual TMA in MAO, mol%	External TMA mmol/L	Activity, kg/(molTi.h) P[a]	St[b]
MAO1	12.8	0	98.7	7.1
MAO1	12.8	20	trace	522
MAO2	28.0	0	trace	783

[a] Propylene polymerization: [Ti] = 0.5 mM; [MAO] = 170 mM; [P] = 0.72 M; T = 60°C; toluene, 20mL; t = 1h. [b] Styrene polymerization: [Ti] = 0.2 mM; [MAO] = 170 mM; [St] = 2.0M; T = 60°C; toluene, 20mL; t = 1h.

However, the catalyst system with MAO1 exhibited a high activity for propylene polymerization. Addition of external TMA into MAO1 system led to an increased activity for styrene polymerization, but deactivated the catalyst for propylene polymerization. Obviously, high content of TMA, involving the residual TMA and the external TMA, in catalyst system favors styrene polymerization, while propylene polymerization could occur only with less TMA. The opposite results from the homopolymerizations suggested that the two monomers are homopolymerized by different catalytic species.

To give a better insight into the effect of residual TMA in MAO on the polymerizations, oxidation state distributions of the Ti species were measured by redox titration. Reaction of the titanocene with MAO1 gave mostly tetravalent Ti (Ti^{+4} = 81.4%), whereas large fractions of Ti were trivalent when the titanocen reacted with MAO2 (Ti^{+3} = 83.3%) or with MAO1 added external TMA (Ti^{+3} = 71.4%). Therefore the tetravalent Ti species can be considered to be active for propylene polymerization and inactive for styrene polymerization. Oppositely, the trivalent Ti species are active for styrene polymerization (20-22), but not for propylene polymerization. It was also indicated that the TMA plays an important role in reduction of the Ti species in the catalyst systems.

Copolymerization of Propylene and Styrene

Copolymerizations of propylene and styrene were conducted with these two catalyst systems and the results are shown in Table II. The composition of the polymerization products is very sensitive to TMA content in the catalyst. As shown in Table II, the polymerizations of a propylene and styrene mixture with catalyst systems containing plenty of TMA, such as the CpTi(OBz)$_3$-MAO2 (run 11) and the MAO1-external TMA catalyst systems (run 8), produced powdered polymer. The products presented ^{13}C NMR spectra displaying only two peaks at 40.9 and 44.1 ppm, respectively, in the aliphatic carbon region and one peak of phenyl C-1 carbons at 145.1 ppm, indicating a syndiotactic polystyrene.

The copolymerizations with CpTi(OBz)$_3$-MAO1 catalyst system gave rubbery polymers (run 3-7). Content of styrene units in the copolymerization products increases with raising the initial concentration of styrene monomer. Propylene homopolymer produced with the catalyst is atactic and easily dissolved in ether, hydrocarbons, and chlorohydrocarbons. The copolymerization products obtained with the catalyst system are completely soluble in heptane except that from run 7 which was obtained in higher styrene concentration and gave a small amount of heptane-insoluble polymer (14 wt%). The heptane-insoluble polymer is a syndiotactic polystyrene as measured by ^{13}C NMR.

Figure 1 shows ^{13}C NMR spectrum of the heptane-soluble fraction of the copolymerization product from run 6 which has a styrene content of 16.3%. In the phenyl C$_1$ carbon region of the spectrum three main peaks appear at 146.8, 146.3 and 144.9 ppm, respectively, which are very different from spectrum of the polystyrene obtained under the same polymerization conditions. The latter shows only one peak at 145.1 ppm in the phenyl C$_1$ carbon region, as shown in the top of Figure 1. The multiple peaks presented by the copolymerization product in this region may be

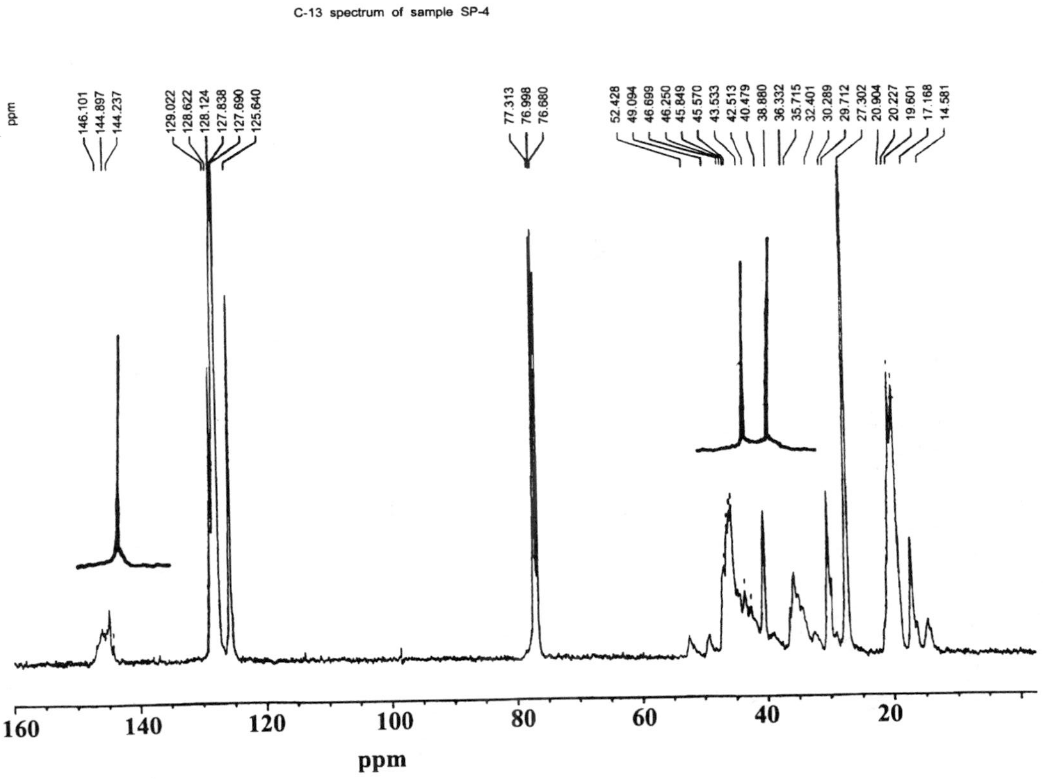

85

C-13 spectrum of sample SP-4

Figure 1. ^{13}C NMR Spectra of Propylene-Styrene Copolymer Prepared with Catalyst System CpTi(OBz)$_3$-MAO1 (from run 6). (For a comparison, the peaks given by styrene homopolymer are inserted in the top).

attributable to styrene units in various copolymeric sequences and tacticities. The peak of methine carbons of styrene units in the copolymerization product moves upfield from 40.9 ppm for the styrene homopolymer to 40.5 ppm. The peak of methylene carbons in styrene homopolymer at 44.1ppm disappears in the spectrum of the copolymer product. No obvious homopolymeric styrene sequences were found in the spectra.

Table II. Results of Copolymerization of Propylene and Styrene in the Presence of CpTi(OBz)$_3$ and different MAO^a

Run	MAO	TMA in MAO, %	External TMA, M	[Pr] M	[St]$_o$ M	Activity kg/(molTi.h)	St in product mol%
1	MAO1	12.8	0	0.72	0	87.7	PP
2	MAO1	12.8	0	0	0.73	5.9	PS
3	MAO1	12.8	0	0.72	0.12	10.2	3.4
4	MAO1	12.8	0	0.72	0.19	9.9	5.8
5	MAO1	12.8	0	0.72	0.38	11.4	9.5
6	MAO1	12.8	0	0.72	0.73	13.8	16.3
7	MAO1	12.8	0	0.72	1.44	10.5	28.1
8	MAO1	12.8	20	0.72	0.73	23.0	PS
9	MAO2	28.0	0	0.72	0	trace	
10	MAO2	28.0	0	0	0.73	130	PS
11	MAO2	28.0	0	0.72	0.38	35.1	PS

^{a} Polymerization: [Ti] = 1 mM; [MAO] = 170 mM; T = 40 °C; toluene, 20mL; t = 1 h.

Polypropylene prepared by the CpTi(OBz)$_3$/MAO1 system is regioirregular (*19*) and contains a nonnegligible proportion of propylene units arranged in head-to-head and tail-to-tail sequences which can be observed from ^{13}C NMR spectrum at 30.3ppm (T$_{\beta\gamma}$), 34.2-35.1ppm (T$_{\alpha\beta}$ and S$_{\alpha\beta}$), 38.3ppm (T$_{\alpha\gamma}$) and 16.2-16.5ppm (Methyl in head-to-head sequence). Incorporation of styrene into polypropylene chains leads to an increase in regioirregularity of the polymer. New peaks can be clearly seen at 35.8, 29.7 and 17.17ppm in the spectra of the present copolymerization products. Intensities of these peaks increase in direct ratio with styrene unit concentration in the copolymer as shown in Figure 2 and Figure 3, respectively.

From a calculation according to the Grant and Paul (*23*) and Randall (*24*) relationships, the peaks at 35.8ppm and 29.7ppm can be individually assigned to the

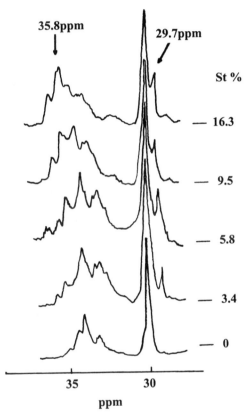

Figure 2. Expanded Spectra of Regioirregular Methylene and Methine Region.

methylene and methine carbons of propylene unit adjacent to a styrene unit in the following tail-to-tail structure.

$$\overset{\displaystyle \text{CH}_3}{\underset{\displaystyle -\underline{\text{CH}}-\underline{\text{CH}_2}-\text{CH}_2-\text{CH}-}{|}}$$

The tail-to-tail structure of Pr-St sequence can be produced by 2,1-insertion of St monomer into the growing chains with 1,2-oriented propylene terminal. The peak at 17.17ppm can be assigned to the methyl carbon of propylene unit adjacent to a styrene unit in a head-to-head mode as the following.

$$\overset{\displaystyle \underline{\text{CH}_3}}{\underset{\displaystyle -\text{CH}-\text{CH}-}{|}}$$

The structure can be prepared by 1,2-insertion of propylene monomer into the growing chains with 2,1-oriented styrene terminal.

It has been suggested that with half-titanocene catalysts propylene polymerization favors 1,2-insertion of the monomer although there is a certain content of chemical inversion (25), while the chain propagation of styrene polymerization occurs *via* 2,1-insertion (6). From such a viewpoint, the regioirregular monomer-alternating sequences, i.e. the St-Pr sequences individually in tail-to-tail and head-to-head enchainments, should be produced when the alternating polymerization reactions (1) and (2) occur.

$$
\begin{array}{ccccc}
\overset{\text{CH}_3}{\underset{|}{}} & & & & \overset{\text{Ph}}{\underset{|}{}} \quad \overset{\text{CH}_3}{\underset{|}{}} \\
\text{Ti-CH}_2\text{-CH----} & + & \text{CH}_2\text{=CHPh} & \rightarrow & \text{Ti-CH-CH}_2\text{-CH}_2\text{-CH----}
\end{array} \qquad (1)
$$

$$
\begin{array}{ccccc}
\overset{\text{Ph}}{\underset{|}{}} & & & & \overset{\text{CH}_3 \ \text{Ph}}{\underset{|\ \ \ |}{}} \\
\text{Ti-CH-CH}_2\text{----} & + & \text{CH}_2\text{=CHCH}_3 & \rightarrow & \text{Ti-CH}_2\text{-CH-CH-CH}_2\text{---}
\end{array} \qquad (2)
$$

The present results are analogous to those from ethylene-styrene copolymers prepared with metallocene/MAO catalysts (13, 26, 27), in which styrene units are isolated and there exists the regioirregular sequence -CH$_2$CH(Ph)CH$_2$CH$_2$CH(Ph)CH$_2$-. From these results, it may be reasonably concluded that real propylene-styrene copolymer can be obtained by the CpTi(OBz)$_3$/MAO1 catalyst system.

Based on the data of the polymer composition (f = d[P]/d[S]) and monomer ratio (F = [P]/[S]$_o$), the monomer reactivity ratios of the propylene-styrene copolymerization catalyzed by the CpTi(OBz)$_3$-MAO1 system were determined according to the method of Fineman-Ross. Figure 4 shows the Fineman-Ross plot,

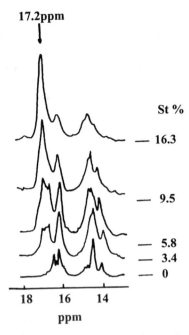

17.2ppm

St %

— 16.3

— 9.5

— 5.8
— 3.4
— 0

18   16   14

ppm

Figure 3.   Expanded Spectra of Regioirregular Methyl Carbon Region.

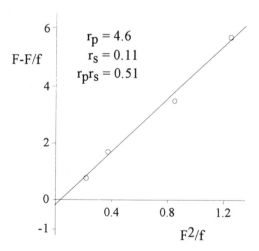

F-F/f

$r_p = 4.6$
$r_s = 0.11$
$r_p r_s = 0.51$

F²/f

Figure 4.   Fineman-Ross Plot for the Copolymerization of Propylene (p) and Styrene (s) with Catalyst System CpTi(OBz)₃-MAO1.

from which the monomer reactivity ratios were obtained to be $r_p = 4.6$, $r_s = 0.11$ and $r_p r_s = 0.51$, indicating a random copolymerization in which styrene presents a reactivity much lower than propylene.

Glass transition temperature (Tg) of the propylene-styrene copolymers was measured by DSC. The values of Tg for the PP and the copolymers containing various amounts of styrene units are shown in Table III. The Tg of the polymer is raised with increasing incorporation of styrene into the polymer chains. The result is consistent with the random structure of the copolymer, as indicated by ^{13}C NMR spectra and the monomer reactivity ratios.

Table III. Glass Transition Temperature of Propylene-Styrene Copolymers

	PP	sample3	sample4	sample5	sample6
St units, mol%	0	3.4	5.8	9.5	16.3
Tg, °C	-8.1	-6.03	-3.3	2.08	12.25

Conclusion

From these results, it can be concluded that propylene-styrene copolymer can be prepared with the catalyst system composed of the half-titanocene and MAO containing less residual TMA. Combining monocyclopentadienyltitanium complex CpTi(OBz)$_3$ with cocatalyst MAO would generate tetravalent Ti species and trivalent Ti species in the ratio depending on the free TMA concentration. The tetravalent Ti species are active for propylene polymerization, and for propylene-styrene copolymerization, but not for styrene homopolymerization. Oppositely, the trivalent Ti species are active for styrene homopolymerization, but inactive for propylene homopolymerization and propylene-styrene copolymerization. These can be summed up in the scheme below. Therefore, the key to success in propylene-styrene copolymerization with the half-titanocene catalysts lies in protection of the metallocene species from reduction.

Acknowledgments
The authors would like to thank the National Natural Science Foundation of China and the Science Foundation of Guangdong Province for financial supports of this research.

References

1. Sinn, H.; Kaminsky, W.; Vollmer, H. J.; Woldt, R. *Angew. Chem., Int. Ed. Engl.* **1980**, *19*, 390.
2. Gupta, V. K.; Satish, S.; Bhardwaj, I. S. *J. Macromol. Sci.,-Rew. Macromol. Chem. Phys.* **1994**, *C34(3)*, 439.
3. Brintzinger, H. H.; Fischer, D.; Mulhaupt, R.; Waymouth, R. M. *Angew. Chem., Int. Ed. Engl.* **1995**, *34*, 1143.
4. Huang, J.; Rempel, G. I. *Prog. Polym. Sci.* **1995**, *20*, 459.
5. Reddy, S. S.; Sivaram, S. *Prog. Polym. Sci.* **1995**, *20*, 309.
6. Po', R.; Cardi, N. *Prog. Polym. Sci.* **1996**, *21*, 47.
7. Bochmann, M. *J. Chem. Soc., Dalton Trans.* **1996**, 255.
8. Zambelli, A.; Ammendola, P.; Proto, A. *Macromolecules* **1989**, *22*, 2126.
9. Oliva, L.; Longo, P.; Grassi, A.; Ammendola, P.; Pellecchia, C. *Makromol. Chem., Rapid Commun.* **1990**, 11, 519.
10. Kakugo, M.; Miyatake, T.; Mizunuma, K. *Catalytic Olefin Polymerization*; Keii, T.; Soga, K., Eds.; Kodansha Ltd.: Tokyo, 1990; pp 517.
11. Longo, P.; Grassi, A.; Oliva, L. *Makromol. Chem.* **1990**, *191*, 2387.
12. Pellecchia, C.; Pappalardo, D.; D扬rco, M.; Zambelli, A. *Macromolecules* **1996**, *29*, 1158.
13. Oliva, L.; Mazza, S.; Longo, P. *Macromol. Chem. Phys.* **1996**, *197*, 3115.
14. Sernetz, F. G.; Mülhaupt, R.; Fokken, S.; Okuda, J. *Macromolecules* **1997**, *30*, 1562.
15. Pellecchia, C.; Proto, A.; Longo, P.; Zambelli, A. *Makromol. Chem., Rapid Commun.* **1992**, *13*, 277.
16. Soga, K.; Yanagihara, H. *Macromolecules* **1989**, *22*, 2875.
17. Soga, K.; Nakatani, H.; Monoi, T. *Macromolecules* **1990**, *23*, 953.
18. Wu, Q.; Ye, Z.; Lin, S. *Macromol. Chem. Phys.* **1997**, *198*, 1823.
19. Wu, Q.; Ye, Z.; Gao, Q.-H.; Lin, S.-A. *J. Polym. Sci: Part A: Polym. Chem.* **1998**, *36*, 2051.
20. Chien, J. C. W.; Salajka, Z.; Dong, S. *Macromolecules* **1992**, *25*, 3199.
21. Longo, P.; Proto, A.; Zambelli, A. *Macromol. Chem. Phys.* **1995**, *196*, 3015.
22. Grassi, A.; Zambelli, A.; Laschi, F. *Orgnometallics* **1996**, *15*, 480.
23. Grant, D. M.; Paul, E. G. *J. Am. Chem. Soc.* **1964**, *86*, 2984.
24. Randall, J. C. *J. Polym. Sci., Polym. Phys. Ed.* **1975**, *13*, 889.
25. Park, J. R.; Shiono, T.; Soga, K. *Macromolecules* **1992**, *25*, 521.
26. Stevens, J. C.; Timmers, F. J.; Wilson, D.R.; Schmidt, G. F.; Nickias, P. N.; Rosen, R. K.; Knight, G. W.; Lai, S. *Eur. Pat.* 416815, **1991**.
27. Oliva, L.; Caporaso, L.; Pellecchia, C.; Zambelli, A. *Macromolecules* **1995**, *28*, 4665.

PROGRESS IN POLYMER DESIGN

Chapter 7

Advantaged Polyethylene Product Design

J. LaMonte Adams, George N. Foster, and Scott H. Wasserman

Univation Technologies, P.O. Box 670, Bound Brook, NJ 08805

Univation Technologies utilizes the combination of Exxpol® metallocene and other emerging catalyst systems adapted to the Unipol® PE process (gas phase polymerization) to provide a pathway for advantaged polyethylene product design. Precise control of the bivariant distribution of molecular weight and branching (or composition), facilitates product design leading to accelerated product development and commercialization. State-of-the art polymer structure and materials rheology measurements yield property descriptors for the sensitive measurement and control of distributed polymer structure of mLPPE. Control of both the molecular weight and comonomer distributions of polyethylene affords the ability to produce polymers with the combination of improved processing properties as well as enhanced product properties including superior toughness and/or excellent optical properties.

Introduction

The global demand for polyethylene (PE) continues at an annual growth rate of nearly 6% per year – e.g., approximately 40 million metric tons of PE demand in 1996 is projected to become 50 million metric tons of PE early in the next decade. The growth for linear-low density polyethylene (LLDPE) is estimated to be about 13% per year.[1] Univation Technologies, the Exxon Chemical/Union Carbide joint venture, was formed in 1997 to license EXXPOL® metallocene catalysts and the UNIPOL® PE process technology. The venture is aligned to develop and license advanced technologies for the manufacture of performance and economically advantaged polyethylene products for high volume markets. In order to provide product technology for the expanding global markets, three types of EXXPOL metallocene catalyzed polyethylene are identified as part of the Univation technology portfolio:

1) **Type I** mLLDPE - narrow molecular weight distribution (MWD), narrow composition distribution (CD), no long chain branching (LCB)

2) **Type II** mLLDPE - broader MWD, narrow to moderate CD, little or no LCB

3) **Type III** mLDPE - narrow to broad MWD, narrow CD, controlled levels of LCB.

Two of the major performance considerations in the polyethylene marketplace are the processability and toughness of products. Other performance considerations may include film optical properties, stiffness/toughness balance, and heat sealing behavior. These performance attributes are strongly influenced by polymer molecular structure and the fabrication process. Through enhanced product design at the molecular level, EXXPOL metallocene technology enables the production of polyethylene products that traverse the toughness-processability domain providing superior properties to those of conventional PE (See Figure 1).

Experimental

State-of-the-art methodologies for structure/property definition are required to describe these new classes of polyolefins. Specific methods used in this paper include dilute solution techniques such as coupled modes of size exclusion chromatography (SEC), temperature rising elution fractionation (TREF), infrared spectroscopy and solution viscometry, or melt rheological experiments such as dynamic oscillatory shear (DOS), steady capillary flow, extensional rheometry and melt tension. Standard density, melt and flow indices (MI, I_2 and FI, I_{21}), and melt flow ratio (MFR = I_{21}/I_2) are routinely measured using ASTM methods.

Molecular weight distributions (MWD) defined in terms of a polydispersity index (PDI = wt. avg. MW/no. avg. MW) were determined by size exclusion chromatography (SEC) with a WATERS 150C GPC instrument equipped with an in-line differential refractometer as the mass detector and a Viscotek viscometer. The coupled SEC-viscometry system operated at 140 °C with 1,2,4-trichlorobenzene as the solvent and mobile phase. Polydisperse polyethylene standards with known MW statistics were used for calibration to obtain the molecular weight and MWD.

Analytical temperature rising elution fractionation (TREF) instrumentation provides a measure of the inter-chain short chain branching distribution (SCBD) or CD.[2] A single beam infrared concentration detector was used to monitor effluent concentration as a function of temperature (inversely related to SCB content).

Another property descriptor that describes the large-scale polymer architecture, the Relaxation Spectrum Index (RSI)[3], was determined from small-amplitude dynamic oscillatory melt rheological tests performed on a Weissenberg Rheogoniometer made by TA Instruments. Controlled rate experiments to calculate the storage and loss moduli were run in parallel plate mode and covered frequencies from 0.1 to 100 sec[-1] at 190°C. Relaxation spectra used to calculate the RSI were generated from the dynamic moduli using commercial rheological software. RSI is defined as the ratio of 2nd to 1st moments of a material's distribution of relaxation times or relaxation spectrum. Capillary and extensional rheological tests were run on custom built instruments at temperatures of 200°C and 150°C, respectively.

The ethylene polymer examples used principally include Type II and III (C2/C6) metallocene low pressure polyethylene (mLPPE) and appropriate Z-N LLDPE and HP-LDPE controls. Additional designations of A and B (*e.g.* Type IIA and Type IIB) indicate different metallocene catalyst families that are capable of producing similar product types.

Results and Discussion

In Figure 2 TREF-CD results are compared for Type I, II, and III mLPPE versus conventional Z-N LLDPE. Overall, the PE produced using EXXPOL metallocene technology have more uniform distributions with less material having comonomer contents at the extremes of the distribution. Conversely, the Z-N LLDPE TREF thermogram is characterized by a pronounced peak located at high temperatures corresponding to high density material (very few short chain branches), as well as a large amount of material that elutes at very low temperature representing low molecular weight, highly branched species. A descriptor known as the CCLDI is used to quantify the CD^4. CCLDI quantifies the distribution of chain segment lengths between branch points (L_i) , defining a crystallizable chain length distribution (CCLD). Analogous to molecular weight polydispersity index (PDI) which is the ratio of two moments of the molecular weight distribution, the CCLDI is a ratio of 2nd to 1st moments of the CCLD, or Lw/Ln. CCLDI values for the most compositionally homogeneous ethylene polymers will be ~1.2-2.5 (narrow CD). Typical values of gas phase Z-N (C2/C6) LLDPE are 12-15 (broad CD), while values for Z-N (C2/C4) LLDPE will be 9-12, and values for EXXPOL mLPPE are 1.5 - 5. The CCLDI statistic is most sensitive to the presence of low SCB or high density fractions. Figure 3 shows the CCLDI-PDI property map for mPE made with Univation technology compared to Z-N LLDPE and LDPE made using the high pressure free-radical polymerization process. The property map shows the wide range of PDI values attainable with EXXPOL mPE technology while maintaining a narrow CD for optimum product performance, thus providing the combination of processability and toughness.

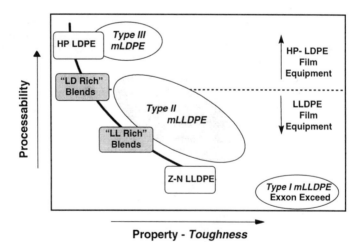

Figure 1. Processability - Toughness Relations for Various PE.

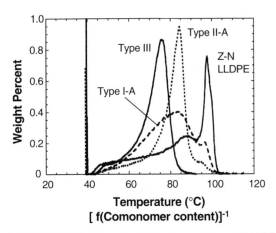

Figure 2. TREF Distributions for Univation mPE and Z-N LLDPE.

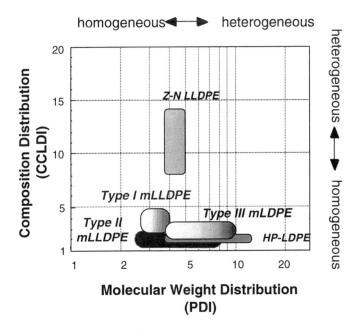

Figure 3. PDI-CCLDI Property Map for Univation mPE and other PE

The RSI has proven to be a sensitive and reliable indicator of long range melt state order (e.g. molecular entanglements), which is indicative of the presence of LCB and/or high MW chains. RSI is a function of polymer MW, MWD, and LCB, as well as temperature. Typical RSI values for 1.0 MI HP-LDPE and Z-N LLDPE are approximately 12-20 and 3-5, respectively. The RSI correlates very well with extrusion ease/maximum output and bubble stability in high rate blown film extrusion[3], as seen in Figures 4 and 5. Both show sensitivity of the RSI to the presence of LCB and high MW species. Comparisons of representative polymer RSI values and the corresponding extruder amperage for blown film processing, are shown in Figure 4. Extruder amperage is one parameter used to measure the relative ease at which polymer melt can be processed. Figure 5 shows the correlation of RSI with melt tension data. Melt tension relates to bubble stability in blown film processing of ethylene polymers.[5] The higher RSI leads to lower required motor load and more easily extruded products.

EXXPOL mLPPEs Blown Film Performance

Figure 6 shows the improved RSI-toughness (dart impact) relationships exhibited by blown films produced from EXXPOL mPE compared to those of Z-N LLDPE, HP-LDPE and blends of the two. This improved combination is achieved via manipulation of the polymer structure on both the local and large-scale molecular levels. Tables I and II show film extrusion data for a developmental Type II mLLDPE product compared with a reference 75%LD/25%LL blend that is typical of what is widely used in the industry. The Z-N (C2/C4) LLDPE grade included in the same table serves as a benchmark for film toughness. The Type II mLLDPE product extrudes as easily as the reference blend, and is accompanied by dramatic improvements in toughness attributes, especially tear strength and dart impact. As expected, it also outperforms the (C2/C4) Z-N LLDPE product by a comfortable margin. This new Type II mLLDPE product will greatly benefit the converters not only in terms of easy processing, but also in down gauging opportunity and the commensurate cost savings. Other EXXPOL Type II mLLDPE technology can provide outstanding blown film optical properties at high toughness with processability equivalent to or greater than Z-N LLDPE.

Table I. Type II Product Processability.

	Z-N (C4) LLDPE	75%/25% LDPE/LLDPE	Type II mLLDPE
Melt Index, I_2 (g/10 min)	1.1	0.6	1.1
MFR (I_{21}/I_2)	24	49	46
Density (g/cm^3)	0.919	0.921	0.919
Output Rate (kg/hr)	50.8	51.7	51.8
Melt Temperature (°C)	221	217	214
Head Pressure (MPa)	23.4	17	15.6
Motor Load (amps)	41	31	31

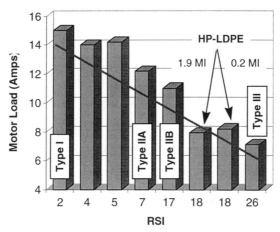

Figure 4. Correlation Between RSI and Extruder Motor Load.

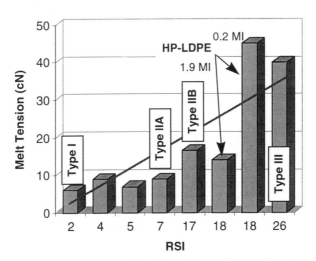

Figure 5. RSI-Processability Correlation with Melt Tension.

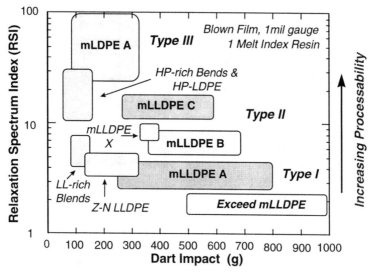

Figure 6. Processability-Toughness Map Cast in Terms of RSI and Dart Impact.

Table II. Type II Film Property Evaluation.

	Z-N (C4) LLDPE	75%/25% LDPE/LLDPE	Type II mLLDPE
Dart Impact Strength (g/25µm)	100	90	410
Elmendorf Tear (g/25µm) (MD)	170	85	210
(TD)	240	140	430
Modulus (MPa) (MD)	203	173	203
(TD)	219	227	215
Film Haze (%)	11.8	8.9	10.2

Type III mLDPEs currently developed by Univation have processability characteristics that are comparable or superior to those of HP-LDPE, yet they possess superior film mechanical properties (Table III and IV). To illustrate, shear viscosity curves for Type III mLDPE, HP-LDPE, and Z-N LLDPE are compared in Figure 7. The shear rheology for the Type III product reasonably matches that of the HP-LDPE, particularly at shear rates typical of blown film extrusion (i.e., 100-1000 sec^{-1}). The combination of physical attributes that are induced by narrow comonomer distribution include enhanced toughness and good optical properties (*e.g.*, low haze and high clarity). Type III mLLDPE's have applications in several areas including uses as general purpose liner, agricultural and shrink films.

Table III compares the resin properties and extrusion characteristics of a HP-LDPE and a representative Type III mLDPE currently in pre-commercial development. The Type III mLDPE shows excellent extrudability and good bubble stability. The head pressure and motor amperage are comparable to those of HP-LDPE of similar MI. No conventional LLDPE could offer such advantaged processability with respect to comparable HP-LDPE. Typical blown film mechanical properties of Type III mLPPE versus a HP-LDPE control are shown in Table IV. Observed are superior toughness/stiffness balance as shown by the tensile-elongation properties at the higher secant modulus. The similar dart impact at higher secant modulus and superior MD/TD tear properties are evident relative to the HP-LDPE control. The mLPPE also show potential for draw-down capability which would broaden their application potential.

Summary

The ever-increasing demands for enhanced polyethylene products have helped foster new developments in polyolefin technology and product design. Through the use of advanced metallocene polymerization catalysts and process technology, Univation Technologies has developed PE technology to address these needs in the forms of Type I, II and III polyethylene. Technology development is focused toward producing metallocene catalyzed low pressure polyethylene film resins with improved

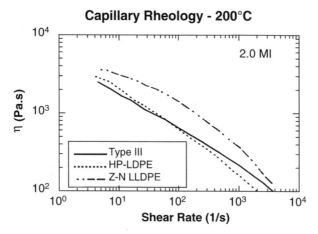

Figure 7. Type III Shear Rheology Comparison.

Table III. Type III Product Processability.

	Type III	HP-LDPE
Melt Index, I_2 (g/10 min)	1.7	1.7
MFR (I_{21}/I_2)	80	51
Density (g/cm^3)	0.922	0.921
Output Rate (kg/hr)	68	68
Screw RPM	25	30
Melt Temperature (°C)	189	191
Head Pressure (MPa)	12.4	12.4
Motor Load (amps)	65	70
Bubble Stability	Good	Good

Table IV. Type III Film Property Evaluation.

		Type III	HP-LDPE
Dart Impact Strength (g/25μm)		82	89
Elmendorf Tear (g/25μm)	(MD)	185	110
	(TD)	320	116
Modulus (MPa)	(MD)	265	179
	(TD)	313	230
Tensile Strength (MPa)	(MD)	24.1	24.5
	(TD)	22.7	19.1
Elongation (%)	(MD)	406	189
	(TD)	606	555

processability and toughness targeted to compete with HP-LDPE in blends and pure structures. State-of-the-art polymer characterization tools and polymer material science methodologies guide product design/development efforts.

Acknowledgments

Technology described in this paper represents the collective efforts and teamwork of numerous catalyst chemists, process engineers, material scientists and product specialists at Exxon Chemical Company, Union Carbide Corporation and Univation Technologies.

References

1. *The Global Polyolefins Industry: 1996, A Period of Structural and Technological Change*, Chem Systems Report, Nov. 1996.

2. Wild, L.; Ryle, T. R.; Knobeloch, D. C.; Peal, I. R., *J. Polym. Sci, Polym. Phys. Ed.*,**1982**, *20*, 441.

3. Wasserman, S. H. *SPE ANTEC Technical Papers,* **1997**, *43*, pp.

4. Foster, G. N. Foster; Wasserman, S. H. "Metallocene And Gas Phase Polymerization: Molecular Engineering Pathway For Advantaged Polyethylene Products", *Metcon '97: Polymers in Transition*, Houston, Texas, 1997.

5. Dealy, J. M.; Wissbrun, K. F. *Melt Rheology and its Role in Polymer Processing,* Van Nostrand Reinhold, New York, 1990.

Chapter 8

Polyolefin Copolymers Containing *p*-Methylstyrene Units: Preparation by Metallocene Catalysts and Application in the Functionalization of Polyolefins

T. C. Chung

**Department of Materials Science and Engineering,
The Pennsylvania State University, University Park, PA 16802**

This paper discusses a new functionalization approach, involving p-methylstyrene (p-MS) "reactive" comonomer. Comparing with other relative styrenic comonomers, i.e. styrene, o-methylstyrene and m-methylstyrene, p-MS shows significantly higher reactivity in metallocene catalysis. The resulting co- and ter-polymers, covering from semicrystalline thermoplastics (PE) to amorphous elastomers (EP and EO), exibit narrow composition and molecular weight distributions. In turn, the incorporated p-MS units in polyolefins, are very versatile, which not only can be interconverted to various functional groups, such as -OH, -NH$_2$, -COOH, anhydride, silane and halides, but also can be conviniently transformed to "stable" anionic initiators for "living" anionic graft-from polymerization reactions. Many new functional polyolefin graft copolymers have been prepared, containing polyolefin backbone (PE, EP, EO, etc.) and functional polymer side chains, such as PMMA, PAN, PS, etc..

Despite the commercial successes of polyolefins (i.e. PE, PP, EP, etc.), the lack of reactive groups in the polymer has limitted many of their end uses, particularly where the interaction with other material is paramount. The logical approach is the use of functionalized polyolefins[1], especially having block and graft structures, as the interfacial compatibilizers[2] to improve interactions in polyolefin blends and composites. Unfortunately, the chemistry to prepare functional polyolefins is very limited both in the direct[3] and post-polymerization[4] processes, namely due to the catalyst poison and the inert nature of polyolefins.

Our functionalization approach have been focusing on the reactive polyolefin intermediates, containing p-methylstyrene[5] (I) and borane[6] (II) reactive groups, as illustrated in Equation 1. By using metallocene technology, both "reactive" comonomers (borane containing α-olefins and p-methylstyrene) can be effectively incorporated into polyolefins. In turn, the reactive comonomer units in polyolefin can be transformed to desirable functional groups under mild reaction conditions. In addition, the reactive comonomer units in polyolefins can also be interconverted to polymeric "living" initiators[7] (free radical in borane case and carbanium in p-MS case) for graft-from polymerization reactions as illustrated in Equation 1. The overall process resembles to the sequential living polymerization, except involving two

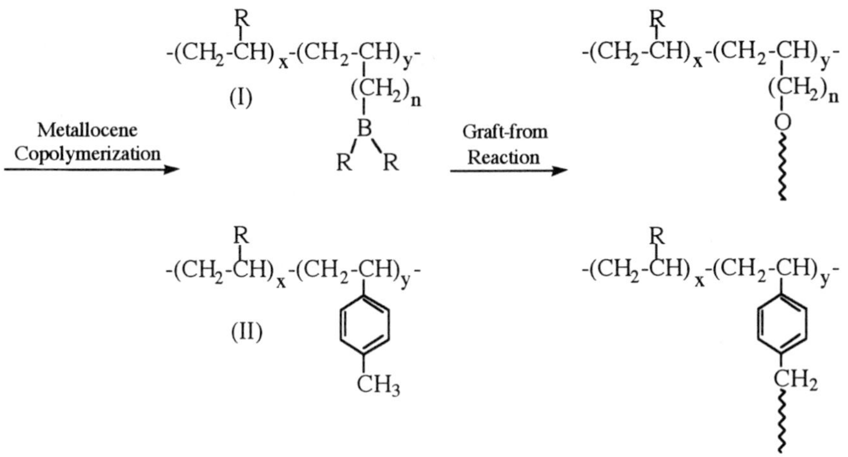

Reactive Copolymers   **Graft Copolymers**

$$-(CH_2-\overset{\overset{\displaystyle R}{|}}{CH})_x-(CH_2-CH)_y- \qquad\qquad -(CH_2-\overset{\overset{\displaystyle R}{|}}{CH})_x-(CH_2-CH)_y-$$

(I) $\overset{\displaystyle (CH_2)_n}{\underset{\overset{\displaystyle B}{R\;\;\;R}}{|}}$ (I) $\overset{\displaystyle (CH_2)_n}{\underset{\displaystyle \sim\!\sim}{\overset{\displaystyle O}{|}}}$

Metallocene Copolymerization →   Graft-from Reaction →

$$-(CH_2-\overset{\overset{\displaystyle R}{|}}{CH})_x-(CH_2-CH)_y- \qquad\qquad -(CH_2-\overset{\overset{\displaystyle R}{|}}{CH})_x-(CH_2-CH)_y-$$

(II) ⬡ CH_3 ⬡ $CH_2\!\sim\!\sim$

Equation 1

polymerization mechanisms. By starting with metallocene polymerization of α-olefins to prepare polyolefin segment, the subsequent functional monomer polymerization is carried out by free radical or anionic mechanisms.

The graft copolymers, containing polyolefin backbone and functional polymer side chains, are not only having high concentration of functional groups but also preserving the original polyolefin properties, such as crystallinity, melting point, glass transition temperature, viscoelaticity, etc.. These segmental polymer structures are known to be the most effective interfacial agent[2] to improve the compatability of polyolefins with other materials, such as glass, metal, fillers and engineering plastics, in polymer blends and composites.

In this chapter, we will focus on the p-MS containing polyolefins, including polyethylene (PE), ethylene-propylene copolymers (EP) and ethylene-1-octene copolymers (EO), with the property range from thermoplastic to elastic. The general polymerization scheme is illustrated in Equation 2.

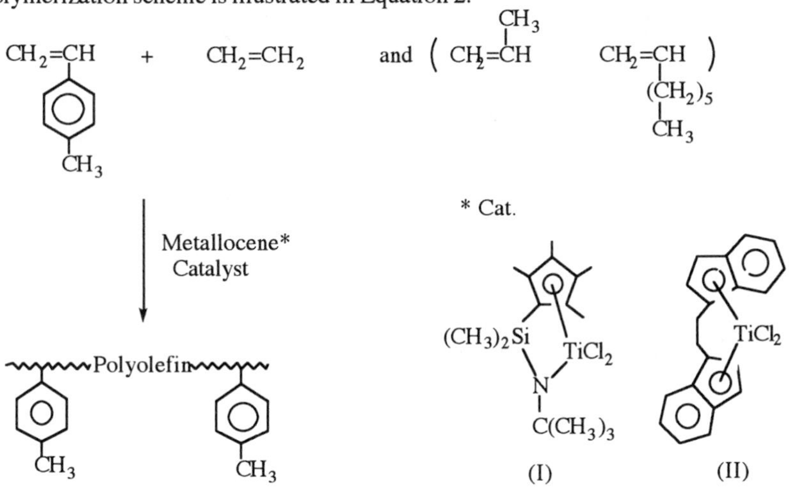

$$CH_2=CH \quad + \quad CH_2=CH_2 \quad \text{and} \quad \left(\underset{\displaystyle CH_3}{CH_2=CH} \quad \underset{\overset{\displaystyle (CH_2)_5}{\underset{\displaystyle CH_3}{|}}}{CH_2=CH} \right)$$

⬡ CH_3

↓ Metallocene* Catalyst

$\sim\!\!\sim\!\!\sim$Polyolefin$\sim\!\!\sim\!\!\sim$

⬡ CH_3 ⬡ CH_3

* Cat.

$(CH_3)_2Si\underset{\overset{\displaystyle N}{\underset{\displaystyle C(CH_3)_3}{|}}}{}TiCl_2$ $TiCl_2$

(I)   (II)

Equation 2

Two metallocene catalysts, [$C_5Me_4(SiMe_2N^tBu)$]TiCl$_2$) (I) and Et(Ind)$_2$ZrCl$_2$ (II), with constrained ligand geometry, were applied in the co- and ter-polymerization reactions. Silicon bridge in catalyst (I) pulls back both Cp and amido ligands from normal positions to form a highly constrained ligand geometry, with Cp-Ti-N angle[8] of 107.6°. On the other hand, ethylene bridge induces the constrained indenyl ligand geometry, with Cp-Zr-Cp angle[9] of 125.8°. Based on the structure-activity relationships of the metallocene catalysts, it is logical to predict that the incorporation of p-MS in catalyst (I), with more opened active site, will be preferable over that of catalyst (II).

Poly(ethylene-co-p-methylstyrene) Copolymers

In a typical copolymerization, the reaction was started by the addition of the metallocene catalyst mixture to a solution of the two monomers in solvent under an inert gas atmosphere. The slurry solution with white precipitates was observed in the reaction. After terminating the reaction with isopropanol, the copolymer was isolated by filtering and washed completely with MeOH and dried under vacuum at 50 °C for 8 hrs. Table 1 summarizes the copolymerization results[10]. The copolymerization efficiency clearly follows the sequence of [$C_5Me_4(SiMe_2N^tBu)$]TiCl$_2$ > Et(Ind)$_2$ZrCl$_2$, which is directly relative to the spatial opening at the active site. In run P-377, about 90% of p-MS was incorporated into copolymer in 1 hour. In run P-383, the reaction produced the copolymer containing 40 mole% of p-MS, which is close to the ideal 50 mole% (the consecutive insertion of p-MS is almost impossible). In general, the catalyst activity systematically increases with the increase of p-MS content, which was also observed in the 1,4-hexadiene copolymerization reactions[11] and could be a physical phenomenon relative to the improvement of monomer diffusion in the lower crystalline copolymer structures. In [$C_5Me_4(SiMe_2N^tBu)$]TiCl$_2$ case, the catalyst activity attains a value of more than 2.4 x 10^6 g of copolymer/mole of Ti x hour in run P-380, which is about 6 times the value for the homopolymerization of ethylene in run P-270 under similar reaction conditions. It is very interesting to note that a very small solvent (hexane and toluene) effect to the catalyst (II) activity was observed in the comparative runs (P-377/P-267) and (P-378/P-379), despite the significant difference in the beginning of reaction conditions (heterogeneous in hexane and homogeneous in toluene). However, the solvent effect is very significant in catalyst (I) systems, hexane solvent conditions consistently show higher p-MS incorporation. The explanation of solvent effect is not clear.

The molecular structure of copolymers were examined by GPC and DSC measurements. Figure 1 compares GPC curves of the polymers prepared by [$C_5Me_4(SiMe_2N^tBu)$]TiCl$_2$ catalyst. The uniform molecular weight distribution in all samples, with $\overline{M}_w/\overline{M}_n$ = 2 - 3, implies the single-site polymerization mechanism. In fact, the GPC curves show a slight reduction of molecular weight distribution in the copolymers, from $\overline{M}_w/\overline{M}_n$ = 2.86 in PE to 1.68 in poly(ethylene-co-p-methylstyrene) containing 18.98 mole% of p-MS. The similar narrow molecular distribution results were also observed in the copolymers prepared by Et(Ind)$_2$ZrCl$_2$ catalyst. The better diffusibility of monomers in the copolymer structures (due to lower crystallinity) may help to provide the ideal polymerization condition. It is very interesting to note that the average molecular weight of copolymers maintain very high throughout the entire composition range, which may be attributed to the relatively high reactivity of p-MS.

Table 1. Summary of The Copolymerization Reactions[b] Between Ethylene and p-Methylstyrene

Run no.	catalyst[a] μmol	ethylene psi	p-MS mol/l	solvent	Temp. °C	yield g	catalyst efficiency kg P/mol M·h	p-ms in copolymer mol%	conversion of p-MS %
p-356	I/17	45	0.085	Hexane	50	6.5	382.4	1.83	47.2
p-372	I/17	45	0.508	Hexane	50	19.8	1164.7	3.94	48.7
p-358	I/17	45	0.678	Hexane	50	21.9	1288.2	5.16	51.1
p-361	I/17	45	1.36	Hexane	50	18.9	1111.8	7.20	29.0
p-371	I/17	45	2.03	Hexane	50	20.8	1223.5	8.94	25.4
p-357	I/17	45	0.085	Toluene	50	5.70	335.3	1.30	29.9
p-392	I/17	45	0.456	Toluene	50	8.63	507.6	3.85	23.2
p-360	I/17	45	0.678	Toluene	50	15.1	888.2	4.76	32.8
p-362	I/17	45	1.36	Toluene	50	19.5	1147.1	6.36	27.0
p-375	I/17	45	2.03	Toluene	50	19.4	1141.2	8.49	22.8
p-270	II/10	45	0	Toluene	30	4.4	440.0	-	-
p-377	II/10	45	0.447	Hexane	30	12.0	1200.0	13.5	90.3
p-378	II/10	45	0.912	Hexane	30	15.5	1550.0	22.6	81.3
p-267	II/10	45	0.447	Toluene	30	13.0	1300.0	10.9	83.8
p-379	II/10	45	0.912	Toluene	30	17.4	1740.0	21.6	86.9
p-380	II/10	45	1.82	Toluene	30	24.2	2420.0	32.8	75.8
p-383	II/10	10	1.82	Hexane	30	15.9	1590.0	40.0	54.6

a. I: Cp_2ZrCl_2/MAO; II: $Et(Ind)_2ZrCl_2$/MAO; III: $[(C_5Me_4)SiMe_2N(t\text{-}Bu)]TiCl_2$/MAO.

b. 45 psi ethylene ~ 0.309mol/l in toluene, 0.424mol/l in hexane at 50 °C; ~ 0.398 mol/l in toluene, 0.523 mol/l in hexane at 30 °C.

10 psi ethylene ~ 0.116 mol/l in hexane at 30 °C.

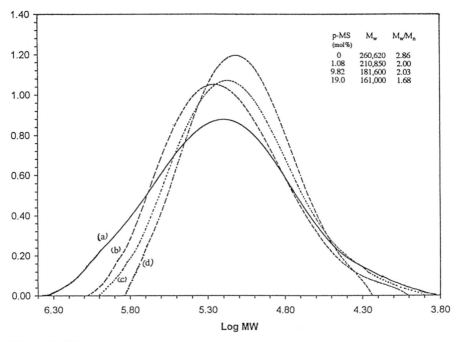

Figure 1. GPC curves of (a) polyethylene and three poly(ethylene-co-p-methylstyrene) copolymers, containing (b) 1.08, (c) 9.82 and (d) 18.98 mole % of p-MS comonomers, prepared by [C$_5$Me$_4$(SiMe$_2$N^{t}Bu)]TiCl$_2$ catalyst, and the molecular weights and molecular weight distributions (insert).

Figure 2 (top) shows the comparison of DSC curves between PE homopolymer and poly(ethylene-co-p-methylstyrene) copolymers prepared by catalyst (I). Even a small amount (~ 1 mole%) of p-MS comonomer incorporation has significant effect to the crystallization of polyethylene. Overall, the melting point (Tm) and crystallinity (χ_c) of copolymer are strongly relative to the density of comonomer, the higher the density the lower the Tm and χ_c. Only a single peak is observed throughout the whole composition range and the melting peak completely disappears at ~ 10 mole% of p-MS concentration. The similar general trend was also observed in the of DSC curves of poly(ethylene-co-p-methylstyrene) copolymers prepared by Et(Ind)$_2$ZrCl$_2$ catalyst. The systematic decrease of Tm and uniform reduction of crystalline curve imply the homogeneous reduction of PE consecutive sequences. The removal of crystallinity provides the opportunity to obtain elastic properties in many thermoplastic polymers. However, the lowest observed Tg is - 5.7 °C in the copolymer containing 18.98 mole% p-MS. With the increase of p-MS, the Tg of copolymer systematically increases as shown in Figure 2 (bottom). The single Tg with sharp thermal transition is indicative of homogeneous copolymer structure. It is clear that poly(ethylene-co-p-methylstyrene) copolymer is difficult to be a good elastomer due to the high Tg of poly(p-methylstyrene). To achieve the elastic polymer, the system requires a third monomer which can provide low Tg. The detail results of elastic poly(ethylene-ter-propylene-ter-p-MS) and poly(ethylene-ter-1-octene-ter-p-MS) terpolymers will be discussed later.

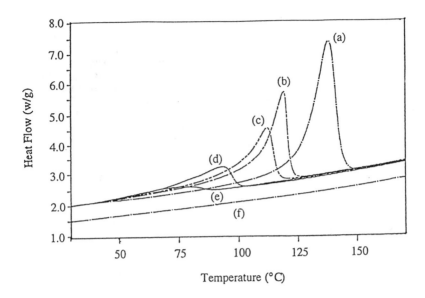

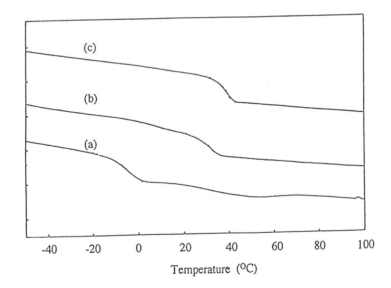

Figure 2. The comparison of DSC curves (top) between (a) polyethylene homopolymer and poly(ethylene-co-p-methylstyrene) copolymers with (b) 1.08, (c) 2.11, (d) 5.40, (e) 9.82 and (f) 18.98 mole% of p-MS comonomer; (bottom) (a) poly(ethylene-co-p-methylstyrene) copolymers with (a) 18.98, (b) 32.8 and (c) 40 mole% of p-MS comonomer.

Comparison Among Styrene Derivatives

It is very interesting to compare p-MS with styrene and methylstyrene isomers[12]. Table 2 compares two sets of copolymerization reactions of ethylene and styrenic comonomers, i.e. p-MS, o-MS, m-MS and styrene, with catalysts (I) and (II), respectively. In both reaction sets, p-MS consistently shows higher incorporation than the corresponding styrenic comonomers, which must due to both favorable electronic and steric effects in p-MS comonomer. The electronic donation of p-methyl group is favorable in the "cationic" polymerization mechanism[13]. On the other hand, the methyl group at para-substitution doesn't effect the monomer insertion. Styrene doesn't have electronic benefit and both isomers (o-MS and m-MS) do not receive the full benefits of the combined electronic and steric effects.

The other way to study the copolymerization reaction is to examine the copolymer molecular structure. Figure 3 compares comparison of DSC curves of poly(ethylene-co-p-methylstyrene) and poly(ethylene-co-styrene) copolymers containing 3 mole% of comonomers, both are prepared by the same $[C_5Me_4(SiMe_2N^tBu)]TiCl_2$ catalyst at 50 °C in hexane.

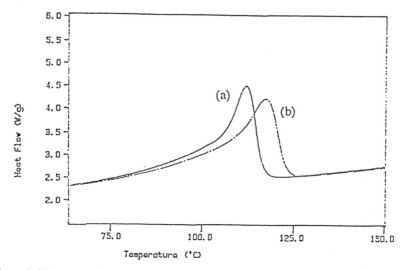

Figure 3. The comparison of DSC curves between (a) poly(ethylene-co-p-methylstyrene) and (b) poly(ethylene-co-styrene) copolymers containing 3 mole% of comonomers.

The significant differences of Tm (111.9 vs. 117.2 °C) imply that p-MS comonomers are more effectively to prevent the crystallization of polyethylene sequences, possibly more uniformly incorporated in copolymer, than styrene comonomers. The similar results were also revealed by ^{13}C NMR measurement. Figure 4 compares the ^{13}C NMR spectrum (with the expanded aliphatic region) of poly(ethylene-co-p-methylstyrene) and poly(ethylene-co-styrene) copolymers containing 10 mole% of comonomers. It is logical to expect that the methyl group substitution at the para-position will have very little effect on the chemical shifts of methylene and methine carbons in the polymer backbone. In general, fewer chemical shifts shown in the poly(ethylene-co-p-methylstyrene) sample imply more homogeneous copolymer microstructure, and every chemical shift can be easily assigned to polyethylene and the isolated p-MS unit. In addition to the two chemical

Table 2. The Comparison of Copolymerization Reactions Between Ethylene (M_1) and Styrene Derivatives (M_2)

Run No.	Reaction Condition[a]			Copolymer Product			
	Temp. (°C)	Cat.[b] (umol)	M_2[c] (mmol)	Yield (g)	M_2 conc. (mol%)	M_2 conv. (%)	Tm (°C)
1	30	(I) 10	none	4.27	0	0	133.7
2	30	(I) 10	p-ms (46.6)	13.0	11.0	84.3	76.0
3	30	(I) 10	o-ms (46.6)	12.9	4.52	38.7	98.3
4	30	(I) 10	m-ms (46.6)	5.43	2.36	9.53	119.1
5	30	(I) 10	styrene (46.6)	13.6	5.35	48.6	98.7
6	53	(II) 17	p-ms (33.9)	24.7	3.30	81.5	114.3
7	53	(II) 17	o-ms (33.9)	23.0	2.54	63.5	118.6
8	53	(II) 17	m-ms (33.9)	19.0	2.35	43.2	118.1
9	53	(II) 17	styrene (33.9)	17.5	1.87	31.7	127.4

a. reaction time = 1 hour, ethylene presure = 45 psi

b. [C$_5$Me$_4$(SiMe$_2$N^{t}Bu)]TiCl$_2$ (I) and Et(Ind)$_2$ZrCl$_2$ (II)

c. p-ms: p-methylstyrene; o-ms: o-methylstyrene; m-ms: m-methylstyrene.

shifts (21.01 and 29.80 ppm), corresponding to the methyl carbon from p-methylstyrene and methylene carbons from ethylene, respectively, there are three well-resolved peaks (27.74, 37.04 and 45.77 ppm) corresponding to methylene and methine carbons from p-MS units which are separated by multiple ethylene units along the polymer chain. On the other hand, the spectrum of poly(ethylene-co-styrene) shows much more complicated methylene and methine carbon species. Many tail-to-tail styrene sequences[8] clearly exist in the polymer chain. The reason for the better seperation of p-methylstyrene units, with no detectable tail-to-tail sequences, may be due to the better regioselectivity of 2,1-insertion of p-MS, with electron donating p-methyl group favorable for "cationic" catalytic site. After the insertion of an ethylene unit (no consective p-MS insertion allowed), the electronic donation of p-methyl group may increase the interaction of the aromatic group with the "cationic" catalytic site to further reduce the space opening around it.

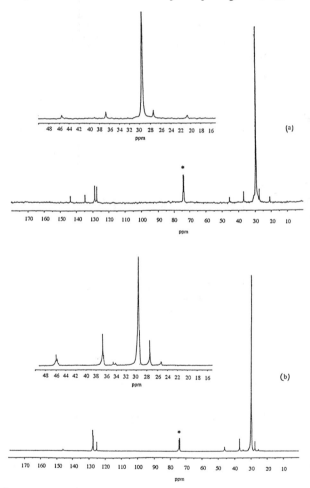

Figure 4, ^{13}C NMR spectra of (a) poly(ethylene-co-p-methylstyrene) with 10.9 mole% p-methylstyrene and (b) poly(ethylene-co-styrene) with 9.5 mole% styrene.

p-MS Containing Polyolefin Elastomers

As discussed in poly(ethylene-co-p-methylstyrene) copolymers, despite the completely amorphous structure the lowest Tg observed in this type copolymer was about - 5 °C, which is too high to be useful in most of elastomer applications. For many commercial applications elastomers with low Tg < -45 °C and having "reactive" sites (such as p-MS units) which can effectively form crosslinking networks and produce stable residues, are very desirable. It is certainly very interesting to expand the poly(ethylene-co-p-methylstyrene) system to polyolefin elastomers. In ethylene-propylene cases, it means preparing a random terpolymer containing close to equal molar ratio of ethylene/propylene and some p-MS "reactive" units. With the unprecedented capability of metallocene technology in copolymerization reactions, it is also very interesting to expand the polyolefin elastomer to new classes containing high α-olefins, such as 1-octene (instead of propylene), which can effectively prevent the crystallization of small consecutive ethylene units and provide low Tg properties. In chemistry, the terpolymerization reaction involving ethylene, 1-octene (high α-olefin) and p-methylstyrene (aromatic olefin) simultaneously, which is very difficult to achieve in Ziegler-Natta polymerization, will be an ultimate test to the metallocene technology.

Poly(ethylene-ter-propylene-ter-p-methylstyrene)

The polymerization reactions were carried out in a Parr reactor. The desirable ratio of ethylene and propylene were mixed in a steel reservoir before piping into the reactor containing a mixed solution of p-MS, MAO and toluene. The polymerization reaction was initiated by charging $[C_5Me_4(SiMe_2N^tBu)]TiCl_2$ catalyst into the monomer mixture. A constant mixed ethylene/propylene pressure was maintained throughout the polymerization process. To assure the constant comonomer ratios, the polymerization was usually terminated in 15 minutes by adding dilute HCl/CH_3OH solution.

Figure 5 compares two GPC curves between poly(ethylene-ter-polypropylene-ter-

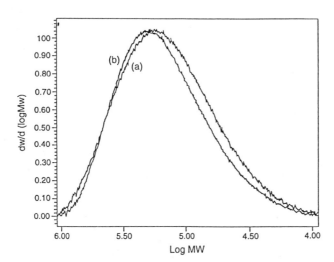

Figure 5. The comparison of two GPC curves of (a) poly(ethylene-ter-polypropylene-ter-p-methylstyrene) terpolymer (run p-120) and (b) poly(ethylene-co-polypropylene) copolymer (run p-116).

p-methylstyrene) terpolymer (with ethylene/propylene ~54/44 mole ratio and 2 mol% p-MS) and the corresponding poly(ethylene-co-polypropylene) copolymer. Both polymers were prepared by $[C_5Me_4(SiMe_2N^tBu)]TiCl_2$/MAO catalyst under similar reaction conditions. Similar molecular weight and molecular weight distribution were observed, which indicate that the addition of p-MS didn't significantly alter the polymerization process (similar propagating rate, no additional termination reaction). The high molecular weight (Mw~ 237,700 and Mn~ 107,500) and narrow molecular weight distribution (Mw/Mn ~ 2.2) is indicative of an ideal single-site polymerization mechanism.

All experimental results[14] with various monomer feed ratios are summarized in Table 3. Overall, the $[C_5Me_4(SiMe_2N^tBu)]TiCl_2$/MAO metallocene catalyst shows excellent activity (3.4-5.4 x 10^6 g polymer/molZr h at 50 °C) in all reactions, with comparative reactivities between ethylene and propylene and good incorporation of p-MS. Such an effective incorporation of an aromatic monomer in polyolefins is very difficult to achieve by traditional Ziegler-Natta catalysts. In the control set (runs p-116, p-117, p-107 and p-115), only involving ethylene and propylene, the copolymer composition is basically governed by the monomer feed ratio. However, with p-MS, the incorporation of propylene seems to slow down. This trend is very clearly observed in the comparative set (runs p-110, p-111 and p-112), with constant ethylene and propylene feeds and various p-MS concentrations; the higher p-MS concentration in the feed the lower propylene/ethylene ratio in the copolymer. As discussed in our previous paper[12], no consecutive p-MS incorporation was observed in $[C_5Me_4(SiMe_2N^tBu)]TiCl_2$/MAO copolymerization reactions due to steric hindrance at the propagating p-MS site. The same steric hindrance may also have some effect on the subsequent propylene insertion, preferring ethylene over propylene. On the other hand, the incorporation of p-MS seems quiet insensitive to the ethylene/propylene feed ratio. In both comparative sets of runs p-120 vs. p-119 and p-128 vs. p-127, with constant p-MS concentration in each comparative run (0.05 and 0.03 mole/l, respectively) and varying ethylene/propylene feed ratio, the incorporation of p-MS is very constant at about 1.6-1.8 and 0.6-0.65 mol%, respectively.

In general, the molecular weight of these terpolymers are very high. Comparing runs p-116 vs. p-120 and p-117 vs. p-119, with the same amount of ethylene and propylene feeds and with and without p-MS, only a small reduction in molecular weight arises from the incorporation of p-MS. It is very interesting to note that the replacement of p-MS with styrene in the same reaction conditions significantly lowers the molecular weight of poly(ethylene-ter-propylene-ter-styrene). The results may be attributed to the relatively comparative reactivity of p-MS with those of ethylene and propylene. The electronic donation of the p-methyl group in p-MS is favorable in this cationic coordination polymerization mechanism.

The glass transition temperature (Tg) was examined by DSC. Figure 6 shows several DSC curves of EP-p-MS terpolymers (samples p-112, p-118 and p-120 in Table 2). Each curve only has a sharp Tg transition in a flat baseline, without any detectable melting point. The combination implies a homogeneous terpolymer microstructure with completely amorphous morphology. The same clean DSC curves were observed in all samples shown in Table 2, even the sample with > 87 mol% of propylene content, which may have most atactic propylene sequences incapable of crystallization. The Tg is clearly a function of the propylene and p-methylstyrene contents. Comparing the ethylene-propylene copolymers (without p-MS units) (runs p-116, p-117, p-107 and p-115), the Tg transitions are linearly proportional to the propylene contents and level off at ~ -50 °C with the composition ~ 50% of propylene content (similar results were reported for the EPDM case[15]). The Tg transition significantly increases with the incorporation of p-MS in ethylene/propylene copolymers. Comparing runs p-117 and p-118, both having ~ 42 mol% of ethylene content and p-MS/propylene mole ratios of 0/58 and 10/48,

Table 3. A Summary of Terpolymerization[a] of Ethylene, Propylene and p-MS by using [(C$_5$Me$_4$)SiMe$_2$N(t-Bu)TiCl$_2$/MAO

Run	Ethylene/Propylene mixing ratio	Monomer Concn. in the Feed (mol/l)			Catalyst Activity	Copolymer Composition (mol%)			Tg	Mv	Mn	PD
No.	(psi/psi)	Ethlyene	Propylene	p-MS	Kgmol⁻¹h⁻¹	[E]	[P]	[p-MS]	(°C)	(g/mol)	(g/mol)	
p116	80/40	0.13	0.28	0	4.9x10³	53.9	46.1	0	-49.4	300444	134285	2.2
p117	70/50	0.12	0.35	0	4.7x10³	41.8	58.2	0	-43.7	284280	120813	2.4
P107	50/50	0.10	0.43	0	4.8x10³	36.2	63.8	0	-35.5	-	-	-
P115	30/70	0.06	0.60	0	4.9x10³	13.0	87.0	0	-16.4	-	-	-
p110	60/60	0.10	0.43	0.1	5.4x10³	37.0	58.5	4.5	-20.3	184149	90634	2.0
p111	60/60	0.10	0.43	0.3	4.9x10³	39.2	52.9	7.9	-19.1	194245	97634	2.0
p112	60/60	0.10	0.43	0.5	5.3x10³	40.3	48.6	11.1	-9.1	188861	80855	2.3
p118	70/50	0.12	0.35	0.3	4.0x10³	46.4	43.6	10.0	-20.7	198199	74890	2.7
p113	50/70	0.08	0.50	0.3	3.5x10³	32.4	59.2	8.4	-12.4	183180	91786	2.0
p120	80/40	0.13	0.28	0.05	4.1x10³	54.4	43.8	1.8	-45.8	237702	107519	2.2
p119	70/50	0.12	0.35	0.05	4.0x10³	46.1	52.3	1.6	-41.0	195760	75028	2.6
p128	85/35	0.14	0.25	0.03	4.4x10³	56.3	43.1	0.6	-48.6	269400	104335	2.6
p127	80/40	0.13	0.28	0.03	3.8x10³	50.7	48.6	0.7	-45.9	244331	85506	2.9

a) polymerization conditions: 100 ml toluene; [Ti]=2.5 x 10⁻⁶ mol; [MAO]/[Ti]=3000; 50°C; 15 min

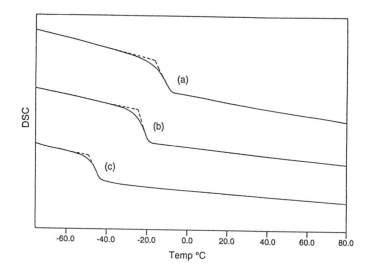

Figure 6. The comparison of DSC curves of EP-p-MS terpolymers from, (a) run p-112, (b) run p-118 and (c) run p-120.

respectively, the Tg increases from -43 to -20 °C. A similar result was observed in the pair of runs p-107 and p-110, with 37 mol% ethylene and smaller difference in p-MS/propylene mole ratios of 0/63 and 4.5/58, respectively, the Tg change is smaller from -35 to -20 °C. It is very interesting to compare runs p-116 and p-120, both with ideal ~ 54 mol% ethylene content and only very small difference in p-MS/propylene mole ratios (0/46 vs. 1.8/44), the Tg's are -50 and -45 °C, respectively. The same Tg trend holds, although much smaller. Overall, the composition of EP-p-MS material with low Tg < -45 °C is very limited, only to the composition with < 2 mol% of p-methylstyrene content. Despite the random terpolymer structure and the ideal (~55/45) ethylene/propylene ratio, further increase of p-MS raises the Tg of the terpolymer to > -40 °C. Obviously, the high Tg's of both propylene (Tg of PP ~0 °C) and p-MS (Tg of poly(p-MS) ~110 °C) components preclude EP-p-MS from achieving some desirable elastomers containing both high content of "reactive" p-MS and low Tg (< -45 °C) transition.

Poly(ethylene-ter-1-octene-ter-p-methylstyrene)

In EP elastomers, the primary function of propylene units is to prevent the crystallization of ethylene sequences. In traditional Ziegler-Natta polymerization, propylene is a natural choice because it has the closest reactivity to ethylene. However, in terms of effectiveness of preventing crystallization of ethylene sequences and obtaining low Tg material, propylene is not the best comonomer, namely due to (i) the small CH_3 side group and (ii) relatively high Tg (~ 0 °C) of the propylene component. In our objective to prepare polyolefin elastomers with low Tg < -45 °C and containing a wide concentration range of "reactive" p-MS units, the EP system clearly shows the serious limitations as discussed above. It is very interesting to replace propylene units with high α-olefins, such as 1-octene, which can effectively prevent the crystallization of ethylene sequences (as known in LLDPE[16]) and is a low Tg material[17] with no possibility of self-crystallization. Additionally, it is also very interesting to study the metallocene technology in a very complicated termonomer system, involving ethylene, 1-octene (high α-olefin) and p-MS (aromatic olefin).

The terpolymerization reaction of ethylene, 1-octene and p-MS was usually started by adding the catalyst mixture of [C$_5$Me$_4$(SiMe$_2$N^{t}Bu)]TiCl$_2$/MAO to the monomer solution, containing 1-octene and p-MS monomers and partially soluble ethylene (with constant pressure) in toluene solvent. Figure 7 compares the GPC curves of the terpolymers prepared under the same ethylene (0.4 mol/L) and 1-octene (0.8 mol/L) and different p-MS (0.1, 0.2 and 0.4 mol/L) concentrations. Overall, the polymer molecular weight is quiet high (Mw ~ 200,000 g/mol) and is not significantly dependent on the content of p-MS. The molecular weight distributions (Mw/Mn) < 2.5, similar to most of metallocene-based homo- and co-polymers, indicate single-site reaction with good comonomer reactivities. The detailed experimental results are summarized in Table 4. Comparing runs p-471/p-470, with similar 1-octene and p-MS concentrations and different ethylene content, the molecular weight of the terpolymer is basically proportional to ethylene content. On the other hand, comparing runs p-471, p-472, p-473, p-474, p-475 and p-476, with the same ethylene concentration and different 1-octene and p-MS ratios, the molecular weight of all terpolymers are very similar with no clear correlation pattern. The molecular weight is clearly governed by ethylene concentration, which must be due to the significantly higher reactivity of ethylene among the three monomers. It is interesting to note that the incorporation of 1-octene is also shown some reduction in high p-MS concentration conditions (comparative runs p-471 vs. p-476 and p-475 vs. p-474). Following the enchainment of p-MS, subsequent insertion of ethylene is faster than that of 1-octene, possibly due to steric hindrance at the active site. In run p-478, using the same comonomer concentration, the resulting terpolymer having ethylene/1-octene/p-MS mole ratio of 9/2/1 clearly indicates the comonomer reactivity sequence of ethylene>1-octene>p-MS. In fact, the ratio is quite consistent with all results (runs p-470, p-471, p-472, p-473, p-474, p-475 and p-476), despite very different comonomer feed ratios.

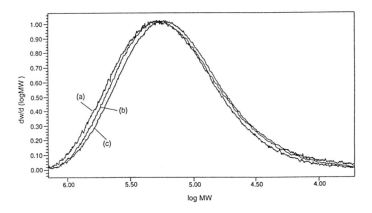

Figure 7. The comparison of GPC curves of three EO-p-MS terpolymers prepared from (a) run p-471, (b) run p-476 and (c) run p-472 .

Table 4. A summary of terpolymerization[a] of ethylene, 1-octene and p-MS using [(C$_5$Me$_4$)SiMe$_2$N(t-Bu)]TiCl$_2$/MAO catalyst.

Run	Monomer concn. In the feed, mol/L			Yield	Copolymer composition mol %			T$_g$	M$_w$	M$_n$	PD
No.	E[d]	1-Oct	p-MS	g	[E]	[O]	[p-MS]	°C	g/mol	g/mol	
p465[b]	0.25	0.89	0	7.0	41.4	58.6	0	-61.8	134,515	60,614	2.2
p466[b]	0.25	0.89	0.13	5.2	40.0	54.5	5.6	-51.3	75,496	36,045	2.1
p470	0.20	0.80	0.10	7.3	54.2	43.0	2.7	-56.2	173,989	74,703	2.3
p471	0.40	0.80	0.10	10.1	61.1	36.0	2.9	-58.1	219,752	96,802	2.3
p472	0.40	0.80	0.20	9.8	60.3	36.3	4.4	-55.7	182,185	77,497	2.4
p473	0.40	0.40	0.20	8.2	59.6	34.0	6.4	-50.1	208,920	86,812	2.4
p477	0.40	0.20	0.20	6.4	80.2	14.1	5.7	-37.3	227,461	91,490	2.5
p476	0.40	0.80	0.40	9.0	63.4	29.3	7.3	-50.3	202,085	96,035	2.1
p475	0.40	0.60	0.40	9.1	67.2	24.7	8.1	-48.2	205,124	93,763	2.2
p478	0.40	0.40	0.40	7.9	73.3	18.5	8.1	-44.7	246,300	122,306	2.0
p474	0.40	0.60	0.15	9.0	64.7	31.3	4.0	-55.7	224,476	102,617	2.2
p396[c]	0.52	0.38	0.91	5.6	64.0	18.1	17.7	-25.8	106,954	55,825	1.9
p383[c]	0.13	0	1.82	15.9	60.0	0	40.0	38.3	-	-	-

a) polymerization conditions (unless otherwise specified): 100 ml of toluene, [Ti] = 2.5 x 10^{-6} mol, [MAO]/[Ti] = 3000, 50 °C, 30 minutes; b) solvent: 100 ml of hexane, 60 °C; c) solvent: 100 ml of hexane, 30 °C, [Ti] = 2.5 x 10^{-5} mol, [MAO]/[Ti] = 2000, ethylene pressure: 45 psi; d) solubility of ethylene: 0.25 mol/L for 29 psi in hexane at 60 °C, 0.20 mol/L for 2 bar, 0.40 mol/L for 4 bar in toluene at 50 °C, 0.52 mol/L for 45 psi in hexane at 30 °C, 0.13 mol/L for 10 psi in hexane at 30 °C. e) solvent: 100 ml of hexane, [Ti] = 10 x 10^{-6} mol, [MAO]/[Ti] = 1500, 30 °C, 60 minutes.

The thermal transition temperature of PO-p-MS terpolymer was examined by DSC studies. Figure 8 shows the DSC curves of two PO-p-MS terpolymers (runs p-472 and p-478) and one poly(ethylene-co-p-methylstyrene) copolymer (run p-383).

Comparing the curves of runs 472 and 383, both having the same ethylene (~ 60 mol%) content but different 1-octene/p-MS ratios, the Tg changes from > 30 °C in run 383 (with no 1-octene) to < -55 °C in run 472 (with 35 mol% 1-octene). The Tg's of co- and ter-polymers are summarized in Table 4. It is very interesting to note that the EO-p-MS sample with even up to 8 mol% of p-MS still shows Tg < -45 °C, which is very different from the corresponding EP-p-MS copolymer as discussed above. These results clearly show the advantages of 1-octene comonomer (over propylene) which assures the formation of amorphous polyolefin elastomer with low Tg and high p-MS content. In Figure 8 (c), the DSC curve of terpolymer (p-478) contains a very weak crystalline peak at ~ 5 °C. Apparently, a total concentration of 1-octene and p-MS more than 30 mol% may be necessary to completely eliminate the crystallization of ethylene sequences.

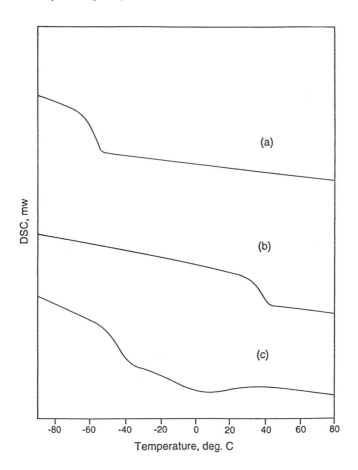

Figure 8. DSC curves of two poly(ethylene-ter-1-octene-ter-p-methylstyrene) terpolymers prepared from (a) run p-472 and (b) run p-478 and (c) poly(ethylene-co-p-methylstyrene) (run p-383).

Anionic Functionalization Reactions

Our major research interest of incorporating p-MS into polyolefins is due to its versatility to access a broad range of functional groups. The benzylic protons are ready for many chemical reactions, such as halogenation, oxidation and metallation as shown in Equation 3. The metallated polymer can further be used in living anionic graft-from reaction, which offers a relatively simple process in the preparation of polyolefin graft copolymers. Usually, the lithiated polymer was suspended in an inert organic diluent before addition of monomers, such as styrene, MMA, vinyl acetate, acrylonitrile and p-methylstyrene. It is interesting to note that the heterogeneous condition allows the easy removal of excess reagent, which was impossible in the homogeneous conditions, such as in the case of poly(isobutylene-co-p-methylstyrene) [18]. The unreacted alkyl lithium complex is much more reactive than benzylic lithium and can produce a lot of undesirable ungrafted homopolymers during the graft-from reaction.

Equation 3

To study the efficiency of lithiation reaction, some of the lithiated polymer was converted to organosilane containing polymer by reacting with chlorotrimethylsilane. Figure 9 compares the ^1H NMR spectra of the starting P[E-co-(p-MS)], containing 0.9 mole % of p-MS, and the resulting trimethylsilane containing PE copolymers, which had been metallated by either s- or n-BuLi/TMEDA, respectively, under the same reaction conditions. In Figure 9 (a), in addition to the major chemical shift at 1.35 ppm, corresponding to CH_2, there are three minor chemical shifts around 2.35, 2.5 and 7.0-7.3 ppm, corresponding to CH_3, CH and aromatic protons in p-MS units, respectively. After the functionalization reaction, Figures 9 (b) and (c) show the reduction of peak intensity at 2.35 ppm and no detectable intensity change at both 2.5 and 7.0-7.3 ppm chemical shifts. In addition, two new peaks at 0.05 and 2.1 ppm, corresponding to Si-$(CH_3)_3$ and ϕ-CH_2-Si, are observed. Overall, the results indicate a "clean" and selective metallation reaction at p-methyl group. The integrated intensity ratio between the chemical shift at 0.05 ppm and the chemical shifts between 7.0 and 7.3 ppm and the number of protons both chemical shifts represent determines the efficiency of metallation reaction. The n-butyllithium/TMEDA converted only 24 mole % of p-MS to benzyllithium. On the other hand, the s-butyllithium/TMEDA was much more effective, achieving 67 mole % conversion. Apparently, the metallation reaction was not inhibited by the insolubility of polyethylene, most of p-MS units must be located in the amorphous phases which are swellable by the appropriate solvent during the reaction. Both DSC and GPC studies, by comparing copolymer samples before and after functionalization reaction, indicate no significant change in the melting point and the molecular weight, respectively.

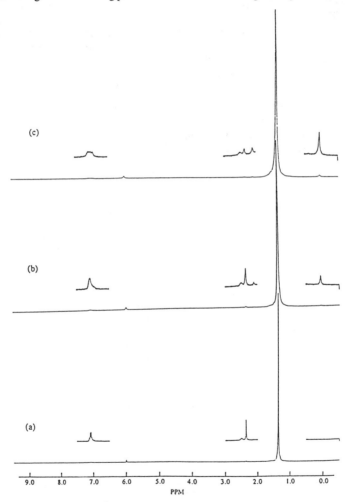

Figure 9. The comparison of ^{1}H NMR spectra of (a) poly(ethylene-co-p-methylstyrene) with 0.9 mole% of p-methylstyrene and two corresponding trimethylsilyl derivatives prepared via lithiation reactions using (b) n-BuLi/TMEDA and (c) s-BuLi/TMEDA reagents.

Anionic Graft-from Reactions

Most of the lithiated PE powder was suspended in cyclohexane before addition of styrene or p-methylstyrene monomers. The living anionic polymerization took place at room temperature, similar to the well-known solution anionic polymerization[19]. To assure sufficient time for monomer diffusion in the heterogeneous condition, the reaction continued for a hour before terminating by the addition of methanol. The conversion of monomers (estimated from the yield of graft copolymer) was almost quantitative (> 90%) in one hour. The reaction mixture was usually subjected to a vigorous extraction process, by refluxing THF through the sample in a Soxhlet extractor for 24 hours, to remove any polystyrene or poly(p-methylstyrene) homopolymers. In all cases, only small amount (< 10%) of THF soluble fraction was obtained. The THF-insoluble fraction is mainly PE graft copolymer and is completely soluble in xylene at elevated temperature. Figure 10 shows the ^{1}H NMR spectra of three PE-g-PS copolymers. Compared with the ^{1}H NMR spectrum of the starting P[E-co-(p-MS)], three additional chemical shifts arise at 1.55, 2.0 and 6.4-7.3 ppm, corresponding to CH_2, CH and aromatic protons in polystyrene. The quantitative analysis of copolymer composition was calculated by the ratio of two integrated intensities between aromatic protons (δ = 6.4-7.3 ppm) in PS side chains and methylene protons (δ = 1.35-1.55 ppm) and the number of protons both chemical shifts represent. Figure 10 (a), (b) and (c) indicate 25.6, 38.1 and 43.8 mole % of PS, respectively, in PE-g-PS copolymers.

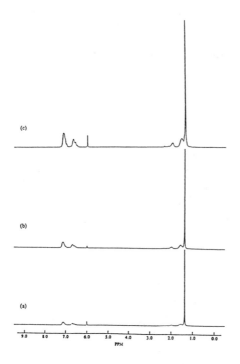

Figure 10. The ^{1}H NMR spectra of PE-g-PS copolymers, containing (a) 25.6 (b) 38.1 and (c) 43.8 mole % of polystyrene.

It is very interesting to note that the same anionic graft-from polymerization can be extended to polar monomers, such as methyl methacrylate (MMA) and acrylonitrile (AN). The reactions take place at room temperature without causing any detectable side reactions as usually shown in the transitional anionic polymerization initiated by butyllithium at ambient temperature. Apparently, the stable benzylic anion may prevent the undesirable addition reaction on the polar groups. Table 5 summarizes the reaction conditions and the experimental results[20]. Overall, the experimental results clearly show a new class of PE graft copolymers, containing either hydrocarbon or polar grafts, which can be conveniently prepared by the tranformation of metallocene catalysis to anionic graft-from polymerization via p-MS groups. The "living" anionic graft-from reaction provides the control to achieve PE graft copolymers with high concentration and high molecular weight of grafted side chains.

Table 5. A Summary of Polyethylene Graft Copolymers

Lithiated PE (g)	Monomer (g)	Solvent	Reaction Temp.(^{o}C)	Reaction Time (hr.)	Isolated Polymer (g)	Graft (mole%)
1.5	ST/1.9	hexane	25	1	3.3	24.4
1.2	ST/5.9	hexane	25	1	6.8	54.7
1.0	p-MS/4.0	hexane	25	0.5	5.0	48.7
1.0	MMA/3.7	THF	0	1.5	1.86	20.0
1.0	MMA/3.4	THF	0	15	2.66	31.8
0.8	MMA/4.0	hexane	25	5	3.08	44.4
0.8	MMA/4.0	hexane	0	5	2.21	33.0
1.0	AN/3.0	hexane	25	16	2.99	51.2

ST: styrene, MMA: methyl methacrylate, AN: acrylonitrile, p-MS: p-methylstyrene

PE/PS Polymer Blends

It is interesting to study the compatibility of PE-g-PS copolymer in HDPE and PS blends. Polarized optical microscope and the SEM were used to examined the surfaces and bulk morphologies, respectively. Two blends comprised of overall 50/50 weight ratio of PE and PS, one is a simple mixture of 50/50 between HDPE and PS and the other is 45/45/10 weight ratio of HDPE, PS, and PE-g-PS (containing 50 mole % of PS). Figure 11 compares the polarized optical micrographs of two blends which were prepared by casting chlorobenzene solutions of the polymer mixtures on glass slides. The optical patterns are very different. A gross phase separation in Figure 11 (a) shows the spherulitic PE and the amorphous PS phases. The PS phases vary widely in both size and shape due to the lack of interaction with the PE matrix. On the other hand, the continuous crystalline phase in Figure 11 (b) shows the compatibilized blend. Basically, the large phase separated PS domains are now dispersed into the inter-spherulite regions and cannot be resolved by the resolution of the optical microscope. The graft copolymer behaving as a polymeric emulsifier increases the interfacial interaction between the PE crystalline and the PS amorphous regions to reduce the domain sizes.

Figure 12 shows the SEM micrographs, operating with secondary electron imaging, which show the surface topography of cold fractured film edges. The films were cryo-fractured in liquid N_2 to obtain an undistorted view representative of the

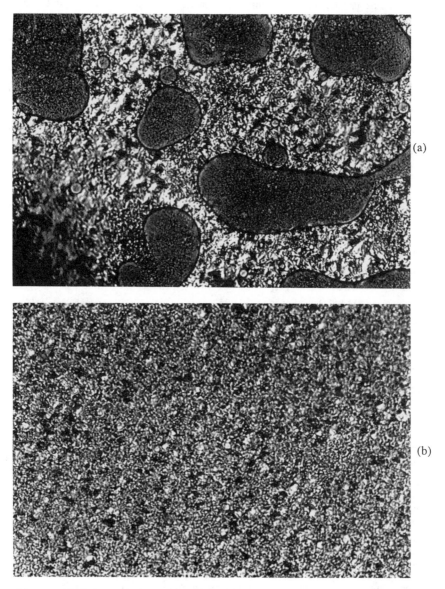

(a)

(b)

Figure 11. Polarized optical micrographs of polymer blends, (a) two homopolymer blend with PE/PS = 50/50 (100x), (b) two homopolymers and PE-g-PS copolymer blend with PE/PE-g-PS/PS = 45/10/45 (100x).

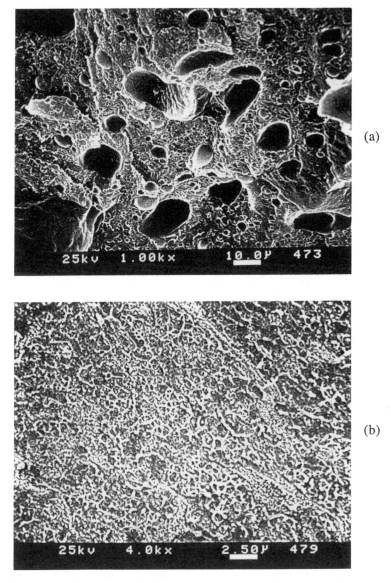

Figure 12.SEM micrographs of the cross-section of two polymer blends (a) two
homopolymers with PE/PS =50/50 (1,000x) (b) two homopolymers and
PE-g-PS copolymer blend with PE/PE-g-PS/PS = 45/10/45 (4,000x).

bulk material. In the homopolymer blend, the polymers are grossly phase separated as can be seen by the PS component which exhibits non-uniform, poorly dispersed domains and voids at the fracture surface as shown in Figure 12 (a). This "ball and socket" topography is indicative of poor interfacial adhesion between the PE and PS domains and represents PS domains that are pulled out of the PE matrix. Such pull out indicates that limited stress transfer takes place between phases during fracture. The similar blend containing graft copolymer shows a totally different morphology in Figure 12 (b). The material exhibits flat mesa-like regions similar to pure PE. No distinct PS phases are observable indicating that fracture occurred through both phases or that the PS phase domains are too small to be observed. The PE-g-PS is clearly proven to be an effective compatibilizer in PE/PS blends.

Conclusion

A new class of reactive polyolefin co- and ter-polymers, containing p-methylstyrene groups, have been prepared by metallocene catalysts with constrain ligand geometry. The combination of spatially opened catalytic site and cationic coordination mechanism in metallocene catalyst provides a very favorable reaction condition for p-methylstyrene incorporation to obtain high polyolefin co-and ter-polymers with narrow molecular weight and composition distributions. The experimental results clearly show that p-methylstyrene performs distinctively better than styrene, o-methylstyrene and m-methylstyrene, in the ethylene copolymerization reaction. In turn, the copolymers are very useful intermediates in the preparation of functional polyolefins and graft copolymers which can serve as the effective compatibilizers in polyolefin blends.

Acknowledgment

Authors would like to thank the Polymer Program of the National Science Foundation for financial support.

Literature Cited

1. (a) Baijal, M. D. *Plastics Polymer Science and Technology*, John Wily & Sons: New York, 1982; (b) Boor Jr., J.*Ziegler-Natta Catalysts and Polymerizations*, Academic Press: New York, 1979; (c). Pinazzi, C., Guillaume, P. and Reyx, D.*J. Eur. Polym.*, **13**, 711 (1977).
2. (a) Riess, G., Periard, J., Bonderet, A.*Colloidal and Morphological Behavior of Block and Graft Copolymers*, Plenum: New York, 1971; (b) Epstein, B. U. S. Patents 4,174,358, (1979); (c) Lohse, D., Datta, S., Kresge, E. *Macromolecules* **24**, 561 (1991).
3. (a) Giannini, U., Bruckner, G., Pellino, E. and Cassata, A. *J. Polym. Sci.: Part C*, **22**, 157 (1968); (b) Purgett, M. D and Vogl, O. *J. Polym. Sci.: Part A: Polym. Chem.* **27**, 2051, 1989; (c) Sivak, A. J. U. S. Patent 5,373,061 (1994). (d) Kesti, M. R., Coates, G. W. and Waymouth, R. M.*J. Am. Chem. Soc.*, **114**, 9679 (1992).
4. (a) Gaylord, N. G. and Maiti, S. *J. Polym. Sci.* **B11**, 253 (1973); (b) Ruggeri, G., Aglietto, M., Petragnani, A.and Ciardelli, F. *Eur. Polymer J.*, **19**, 863 (1983); (c) Michel, A. and Monnet, C. *Eur. Polym. J.*, **17**, 1145 (1981).
5. Chung, T. C. and Lu, H. L. U. S. Pat. 5,543,484 (1996)

6. (a) Chung, T. C., Jiang, G. J. and Rhubright, D. U. S. Patents 5,286,800 (1994);
   (b) Chung, T. C., Jiang, G. J. and Rhubright, R. U. S. Patents 5,401,805 (1995);
   (c) Chung, T. C. and Rhubright, D.*Macromolecules*, **26**, 3019 (1993); (d) Chung,
   T. C., Lu, H. L. and Li, C. L.*Polymer International*, **37**, 197 (1995).
7. (a) Chung, T. C., Lu, H. L. and Janvikul, W.*J. Am. Chem. Soc.*, **118**, 705
   (1996); (b) Chung, T. C. and Jiang, G. J.*Macromolecules*, **25**, 4816 (1992); (c)
   Chung, T. C., Rhubright, D. and Jiang, G. J. *Macromolecules*, **26**, 3467 (1993).
   (d) Chung, T. C. and Rhubright, D.*Macromolecules*, **27**, 1313 (1994); (e) Chung,
   T. C., Janvikul, W., Bernard, R. and Jiang, G. J. *Macromolecules*, **27**, 26 (1994);
   (f)) Chung, T. C., Janvikul, W., Bernard, R., Hu, R., Li, R. C., Liu, S. L. and
   Jiang, G. J. *Polymer*, **36**, 3565, (1995).
8. Stevens, J. C. *Stud. Surf. Sci. Catal.*, **89**, 277 (1994).
9. Ewen, J. A., Jones, R. L., Razavi, A. and Ferrara, J. L. *J. Am. Chem. Soc.*, **110**,
   6255 (1988).
10. Chung, T. C. and Lu, H. L.*J. of Polym. Sci. A, Polym. Chem. Ed.*, **35**, 575
    (1997).
11. Chung, T. C., Lu, H. L. and Li, C. L.*Macromolecules*, **27**, 7533 (1994).
12. Chung, T. C. and Lu, H. L.*J. of Polym. Sci. A, Polym. Chem. Ed.*, **36**, 1017
    (1998).
13. (a) Jordan, R. F. *J. Chem. Edu.*, **65**, 285 (1988). (b) Eshuis, J. J., Tan, Y. Y.,
    Meetsma, A. and Teuben, J. H.*Organometallics*, **11**, 362 (1992). (c) Yang, X.,
    Stern, C. L. and Marks, T. J. *J. Am. Chem. Soc.*, **116**, 10015 (1994).
14. Lu, H., S. Hong and Chung, T. C.*Macromolecules*, **31**, 2028 (1998).
15. (a) Baldwin, F. P., Ver Strate, G. *Rubber Chem. Technol.*, **45(3)**, 709 (1972). (b)
    Ver Strate, G. *Encycl. of Polym. Sci. and Eng.*, **6**, 522 (1986).
16. Canich, J. M. U.S. Pat. 5,026,798 (1991).
17. Plate, N. A., Shibaev, V. P. *J. Polym. Sci.: Macromol. Rev.*, **8**, 117 (1974).
18. Powers, K. W., Wang, H. C., Chung, T. C., Dias, A. J., Olkusz, J. A. U. S.
    Pat. 5,548,029 (1996).
19. Szwarc, M. *Adv. Polym. Sci.*, **47**, 1 (1982).
20. Chung, T. C., Lu, H. L. and R.D. Ding.*Macromolecules*, **30**, 1272 (1997).

PROGRESS IN CHARACTERIZATION

Chapter 9

The Distribution of Asymmetric and Symmetric Chains in Highly Isotactic Polypropylenes

J. C. Randall[1], C. J. Ruff[1], and Masatoshi Ohkura[2]

[1]Baytown Polymers Center, Exxon Chemical Company, Baytown, TX 77522
[2]Tonen Chemical Corporation, Technical Development Center 3-1,
Chidori-Cho, Kawasaki-Ku, Kawasaki 210, Japan

Structural distributions have been determined for the components of an impact copolymer, which are prepared in series reactors with the same magnesium chloride supported titanium based Ziegler Natta catalyst system. The structural distribution of a polypropylene homopolymer prepared in the first reactor is compared to the structural distribution of a downstream E/P copolymer to ascertain if this information is revealing about the catalyst site distribution. The amorphous and crystalline components of the second stage E/P copolymer exhibit completely different E/P triad sequence distributions, which suggest they originate from different types of catalyst sites. The homopolymer asymmetric/symmetric chain structural distribution obtained from a Doi two state statistical fit of the pentad/heptad *meso,racemic* sequence distributions is compared to the ratio of crystalline to amorphous E/P copolymers prepared in the second stage. A distribution of 80/20 for asymmetric versus symmetric chains satisfactorily fits the Doi two state homopolymer statistical model and accounts for the structural distributions observed for the E/P copolymer ICP components.

It has long been recognized that heterogeneous titanium based Ziegler-Natta catalysts are multi-sited(1). Structurally unique polypropylene molecules are not produced during polymerization; instead polypropylene molecules that differ in levels of stereo- and regio-irregularities and subsequently in crystallinities are made simultaneously. The clearest example of this type of catalytic behavior is the presence of atactic polypropylenes as a component within bulk, highly isotactic polypropylenes. The crystalline polypropylenes, which accompany the amorphous, atactic molecules, vary in levels of crystallinity (1,2,3) and exhibit peak melting points and breadths of curves that reflect the level and distribution of stereo- and regio-irregularities(3). It is the purpose of this study to add to the understanding of the structural distributions that occur in isotactic polypropylenes prepared with classical Ziegler-Natta catalysts. Toward this end, an "impact copolymer"(ICP), where the same catalyst system is used throughout a sequential polymerization, was also examined for its structural distribution. Impact copolymers are synthesized in two steps where a polypropylene homopolymer is prepared initially and an ethylene/propylene copolymer is made in a second stage. The combined lengths of the two polymerizations require that the single catalyst system remain active throughout the duration of the sequential polymerization process. These types of sequential polymerizations(4), lead to a family of thermoplastic olefins called high impact polypropylenes(ICPs), which are finding increasing applications in the consumer, appliance and automotive industries. These in situ blends consist of a polypropylene homopolymer matrix with the lesser amounts of immiscible ethylene propylene copolymer components existing typically as small, globular particles dispersed throughout the homopolymer matrix(4,5). It is interesting to establish if the ethylene/propylene copolymers prepared with the same catalyst system exhibit a structural distribution analogous to the first stage polypropylene homopolymers.

From a stereoregularity viewpoint, isotactic polypropylenes can be viewed as having either or both of two types of configurational sequences(6): asymmetric chains, where a particular configuration strongly dominates, and symmetric chains, where the populations of individual repeat unit configurations are equivalent. Asymmetric chains are often called "enantiomorphic-site control" polymers (7), which reflect the stereospecificity of the catalyst and symmetric chains are often called "chain-end" control polymers to imply no configurational preference by the catalyst (8,9). From a statistical standpoint, both symmetric and asymmetric chain sequence distributions have been treated successfully by Bernoullian statistics, which utilize a single parameter to describe transitions between Markovian states. In terms of the Price 0,1 nomenclature(6), asymmetric chains require that

$$P_{00} \gg P_{11}$$

and

$$P_{10} \gg P_{01}$$

or

$$P_0 \gg P_1.$$

Symmetric chains require that

$$P_{00} = P_{11} = P_m.$$

and

$$P_{10} = P_{01} = P_r,$$

which is a special case of first order Markovian statistics(10).

It should be noted that Markovian statistics are used to describe polymer chain sequence distributions as opposed to stipulating that the sequence distributions represent a particular type of catalyst behavior, that is, whether enantiomorphic site control (P_0) or chain end control (P_m). There is some danger in describing polymer configurational sequences too broadly by catalyst behavior. For example, with some metallocene catalysts, it is possible to produce a symmetric chain from an enantiomorphic site control catalyst through the proposed alternating mechanism(11). Over 25 years ago, F. P. Price(6) noted, "Markovian mathematics does not necessarily demand particular reaction schemes for the production of the polymer. The mathematics is only a framework within which it is possible to describe polymer chains having particular sequential characteristics, regardless of how these chains were produced." In studies of polypropylene structures produced by different types of catalyst systems, it should be noted whether symmetric chains, asymmetric chains or a mixture of the two types are produced. Isotactic polypropylenes, produced with classical Ziegler-Natta catalysts, offer examples where both symmetric and asymmetric chains can be produced simultaneously at different catalyst sites.

There have been a number of attempts to decipher asymmetric/symmetric chain distributions. Doi(12) and Hayashi, et al.,(13) used two state, zero order Markovian statistical models for both symmetric and asymmetric chains to mimic the distribution of polypropylene stereochemical structures obtained during Ziegler-Natta polymerizations. Coleman and Fox(14) used a two-state model to take into account reversible switches between catalyst sites producing different compositions of symmetric chains. Cheng(15) has discussed the various possible combinations of two state models that can be used to simulate observed polypropylene sequence distributions. In a study analogous to this investigation, Chûjo, et al.,(16) used a two state model to fit sequence distribution data from a series of polypropylenes created with different electron donors and suggested that some catalyst sites could fluctuate between producing asymmetric and symmetric chains. Busico, et al.(17), introduced new parameters into a Bernoullian two state model to account for syndiotactic blocks in highly isotactic polypropylenes. Recently, it was shown that a two state model using first order Markovian symmetric chain statistics in place of Bernoullian symmetric chain statistics will lead to a prediction of short syndiotactic blocks in highly isotactic polypropylenes(18).

The crystallizable components from a highly isotactic polypropylene will generally have an isotactic pentad fraction [$mmmm$] of at least 0.97, a $meso$ diad content of at least 0.98 and a distribution where asymmetric chains dominate over symmetric chains. Probably the best model for attempting to determine the symmetric/asymmetric chain distribution is the Doi 2-state model(12). In this study, the Doi 2-state model, used successfully by Doi(12), Bovey(9), and Shelden, et al.(7), and others on polypropylenes with lower isotactic diad levels, will be employed with Bernoullian statistics for the asymmetric chain component and first order Markovian statistics for the symmetric chain component(18). The use of Bernoullian statistics in the Doi two state model for the symmetric chain component leads to what appears to be unreasonably large differences between the contents of asymmetric versus symmetric sequence distributions for highly isotactic polypropylenes(16,17), as will be demonstated shortly. The asymmetric chain transition probability, P_o, (called σ by Doi), is greater than 0.99 for asymmetric chains while P_r is around 0.6 for the symmetric chain components. This leads to a 99+ percent asymmetric chains (termed ω by Doi) in the final distribution. The results for the original Doi two-state model were duplicated in this study and are also very close to those reported by Chûjo, et al.(16) and Busico, et al.(17) for corresponding highly isotactic polypropylenes.

IPCs offer an opportunity to gain further insights into the classical Ziegler-Natta catalyst behavior because high impact polypropylenes involve homopolymer/copolymer sequential polymerizations using the same catalyst system. New information about catalyst behavior could be provided from observed similarities and differences among the respective behaviors of the homopolymer stereochemical sequence distributions versus the copolymer E/P sequence distributions. Such an understanding depends upon a detailed knowledge of the structures of the ICP components, which consist of crystalline polypropylene, atactic polypropylene, an amorphous E/P copolymer and a crystalline E/P copolymer(19). In this study, the following ICP structural information was obtained: the fraction and stereoregularity of the crystalline isotactic polypropylene(c-PP), the fraction of atactic polypropylene(a-PP), the fraction and composition of an amorphous ethylene/propylene copolymer(a-E/P) and the fraction and composition of a crystalline ethylene/propylene copolymer(c-E/P)(20). The latter copolymer component falls into the class of linear low density polyethylenes, which are often found in the core of the dispersed phase amorphous copolymer particles of high impact polypropylenes(21).

E/P triad information retrieved from NMR data of the c-E/P and a-E/P copolymers is used to determine an apparent reactivity ratio product, $r_1 r_2$, which is an indicator of whether the two components arise from the same or different catalyst sites(20). An apparent $r_1 r_2$ of unity indicates a narrow composition distribution, a random sequence distribution and likely origination from a unique type of catalyst site(22). A value greater than unity indicates a departure from a narrow composition distribution, which most likely indicates production from a distribution of catalyst sites(22). It was observed in an earlier study of over 100 ICPs, differing by composition and E/P copolymer amounts, that the amorphous E/P components gave apparent reactivity ratio products consistently around unity(20.). The corresponding crystalline E/P components gave apparent reactivity ratio products in a range from around 2 to 25(20). This behavior suggests that the crystalline and amorphous E/P components likely originate from different catalyst sites as opposed to representing different ends of the composition distribution.

Experimental

A sample of polypropylene homopolymer was taken during the first stage of a sequential polymerization of propylene and ethylene/propylene to form an ICP. A magnesium chloride supported fourth generation Ziegler Natta catalyst system was employed with the external donor, DCPMS, (dicyclopentyldimethoxysilane), which is known to produce highly isotactic polypropylenes(23). The homopolymer stereoregularity was characterized by measuring [m], [mmmm] and the average meso run length (MRL) by ^{13}C NMR(18), as shown in Table 1. The ^{13}C NMR spectrum of

Table 1. Stereoregularity of an ICP Homopolymer Component

	PP Component from an ICP
[mmmm]	0.9862
[m]	0.9946
Average meso run length, (MRL)	471
% xylene insolubles	99.09
% xylene solubles	0.87

the homopolymer revealed no detectable resonances from regio-irregularities, although n-butyl end groups were observed. As discussed by Chadwick(24), mis-insertions in highly isotactic catalyst systems such as the one used in this study predominantly undergo chain transfer to form n-butyl end groups and are seldom seen as enchained regio defects. Consequently, the study could be restricted to a characterization of stereo irregularities, detected at levels of ~0.0001 mole fraction and higher. Amorphous polypropylenes were removed from the homopolymer sample so that it could be assured that the level of measured stereo-defects were from within a crystalline polypropylene chain. This was accomplished by fractionating the sample into "cold xylene insolubles" and "cold xylene solubles"(20). The xylene soluble fraction was 0.87%, as shown in Table 1. Both fractions were examined by ^{13}C NMR, which revealed the solubles to be a typical atactic polypropylene. Table 1 contains the xylene solubility information.

Two high impact copolymers prepared with the DCPMS donor, ICP-A and ICP-B, were also quantitatively fractionated into xylene solubles and insolubles at 21 °C by Polyhedron Laboratories, Inc.(20)(25). The xylene fractionation data, showing excellent material balances, are given in Table 2a. Each of the xylene fractions were subsequently examined using ^{13}C NMR and characterized according to a previously published procedure(20). Structural data for the ICPs are given in Tables 2a and 2b.

Table 2a. Characterization of Crystalline Polypropylenes in ICPs produced with the DCPMS Electron Donor.

Sample	% xylene sol.	% xylene insol.	Wt. % c-PP	[meso]	Avg. MRL
ICP-A	21.99	78.07	73.60	0.990	399
ICP-B	24.57	75.49	71.69	0.987	394

Table 2b. Compositions of ICPs produced with the DCPMS Electron Donor.

Sample	% a-PP	% a-E/P	% E in a-E/P	% c-E/P	% E in c-E/P
ICP-A	2.80	19.19	41	4.47	82
ICP-B	3.45	21.12	40	3.80	76

A comparison of the directly determined tacticity of the polypropylene homopolymer extracted from the first stage of the sequential polymerization is similar to that indirectly determined for the homopolymer components of ICP-A and ICP-B. An NMR deconvolution procedure was utilized to establish the structural distributions from the final ICP products(20).

Carbon 13 NMR data were obtained at 100 MHz at 125°C on a Varian VXR 400 NMR spectrometer. A 90° pulse, an acquisition time of 4.0 seconds and a pulse delay of 10 seconds were employed. The spectra were broad band decoupled and were acquired without gated decoupling. Similar relaxation times and nuclear Overhauser effects are expected for the methyl resonances of polypropylene(26), which were the only homopolymer resonances used for quantitative purposes. A typical number of transients collected was 2500. The samples were dissolved in tetrachloroethane-d$_2$ at a concentration of 15% by weight. All spectral frequencies were recorded with respect to an internal tetramethylsilane standard. In case of the polypropylene homopolymer, the methyl resonances were recorded with respect to 21.81 ppm for *mmmm*, which is close to the reported literature value of 21.855 ppm(13) for an internal tetramethylsilane standard. The pentad/heptad assignments used in this study are well established(13,27,28).

Spectral integrations were obtained with scale settings between 5000 and 1,000,000, depending upon the strength of the signal being measured. At completion, all resonance areas were converted to a common scale factor. In an earlier study of polypropylene homopolymers(18), integrations were performed by using the same integral limits for each pentad/heptad resonance for all of the samples. In this study, the integral measurements were made with the limits placed as close to each resonance as possible. The most significant effect is an increase in the average *meso* run length for the highly isotactic polypropylene in this study as compared to the earlier study. It is important to review and experiment with the procedure for measuring the resonance areas. Peak heights were also examined because it is necessary to determine precisely the relationships between *mmmr*, *mmrr* and *mmrrmm*. In the homopolymer, the ratio of *mmmr* to *mmrr* to *mmrrmm* is 1.00 to 0.86 to 0.33, which is similar to the earlier reported ratios(18) for a DCPMS donor with the same catalyst system. The *mmrrmm* relative area of 0.33 is considerably less than that expected for a 2:2:1 ratio. The presence of a symmetric chain component will lower the *mmrr* to *mmrr* ratio to a small extent, but it can cause *mmrrmm* to diminish significantly relatively to *mmmr*, *mmrr*. The observed ratios of *mmmr* to *mmrr* to *mmrrmm* are very important when determining the amount of symmetric chains that are present.

The "solver routine" of Excel-4™(29) was used to minimize standard deviations between calculated Markovian and observed methyl pentad/heptad sequence distributions as well as the E/P copolymer triad distributions(18,20). The solver routine is particularly well suited for calculations of this type because it is easy to switch from zero to first order Markovian models when testing the behavior of the observed sequence distributions. Care must be taken to identify the optimum minimum between calculated and observed results because more than one minimum is typically found by the solver routine for any particular pentad/heptad distribution. In case of the copolymer analyses, pentad data for [EPEPE], [PPEPE+EPEPP] and [PPEPP] were also included in the Markovian statistical analyses to increase the number of independent observations for the determination of the statistical parameters.

Symmetric and Asymmetric Polypropylenes

Improvements in polypropylene stereoregularity, obtained by only increasing DCPMS concentrations with the same catalyst system was observed in an earlier study(18). The homopolymer pentad/heptad distribution in this study contains sequences with consecutive *racemic* diads even at the observed high level of stereoregularity ([*m*] = 0.99). See Table 3. It is clear that syndiotactic types of chain defects are present because *racemic*-centered heptad and other pentad resonances are observed in addition to the *mmmr, mmrr* and *mmrrmm* resonances for an isolated, inverted configuration (~00000100000~) found principally in asymmetric chains(18).

The presence of syndiotactic sequences is important because it clearly indicates the presence of crystalline symmetric chains(18) even at very high isotactic levels where the amount of atactic polypropylene is now below 1 per cent (Table 1). Typical structures of crystalline isotactic symmetric chains (17) are,

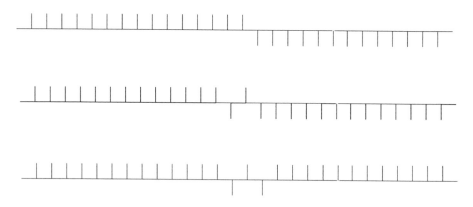

where the vertical lines above point up or down to indicate differences in propylene methyl group configurations. In a previous study, a single, isolated unit with a configuration inverted from that of the main chain, which is characteristic of an asymmetric chain, was the only type of chain defect observed by Paukkeri, et al.,(21). The determination was from a highly isotactic polypropylene fraction with a similar [*mmmm*] content to the polypropylene examined in this study, but obtained from an overall lesser isotactic polypropylene. Paukkeri, et al.,(3) also reported syndiotactic blocks in crystalline polypropylenes at moderate isotactic levels, but again was unable to detect syndiotactic sequences in polypropylene fractions of exceptional isotacticity. Paukkeri's results are expected for those polypropylenes where the symmetric chains exhibit a lower level of isotacticity than do the asymmetric chains. It is anticipated that the highest molecular weight fractions could be purely asymmetric chains. In another related study, Busico, et al., explained the existence of syndiotactic blocks within long isotactic sequences by suggesting that a catalyst site could reversibly switch from "isotactoid" to "syndiotactoid" sequences in times less than the growth times of the polymer molecules(17,28,30). In an earlier stage of this study, it was shown that short syndiotactic blocks connecting longer isotactic blocks could be accurately predicted by employing first order Markovian symmetric chain statistics to describe the sequence distributions (18).

The presence of syndiotactic sequences of at least 3 to 7 units in length in highly isotactic polypropylenes indicates the presence of some level of isotactic symmetric chains. It is highly desirable to be able to determine the distribution between asymmetric and symmetric chains in typical isotactic polypropylenes. One advantage of comparing an ethylene/propylene copolymer to a polypropylene homopolymer synthesized with the same catalyst system is that the copolymer compositional distribution is easier to characterize and it may lead to independent information about the distribution of catalyst sites. The E/P compositional information can be employed in the Doi 2-state homopolymer model to establish the fraction of asymmetric chains. It is then only necessary to optimize the Markovian statistical parameters for the best fit of the observed pentad/heptad distribution.

The Doi two-state 0/0 model(12), which utilizes both Bernoullian symmetric and asymmetric chain statistics and a weighting factor to describe the distribution between the two types of chains, was developed to determine if the fractions of symmetric versus asymmetric chains could be determined by fitting an observed ^{13}C NMR methyl pentad distribution. (The symbol 0/0 is used throughout to describe those

137

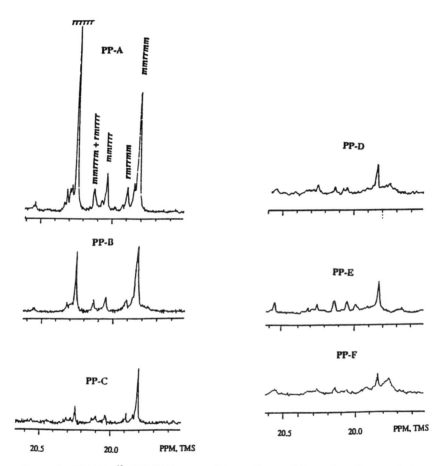

Figure 1. 50 MHz ^{13}C NMR Spectra of the *rr*-Centered Pentads and Heptads for PP-A through PP-F. (Reprinted with permission from *Macromolecules*, February 1997, *30*, pp 803-816. Copyright 1997 American Chemical Society)

models where Bernoullian statistics are used for both asymmetric and symmetric chain components.) Three Doi 0/0 parameters, P_0 where $P_0 > P_1$, (called σ by Doi), P_m where $P_{00} = P_{11}$ (Doi used 1- P_m or P_r) and ω, the weight factor, were used to fit observed polypropylene pentad distributions. At low isotactic levels, Bernoullian statistics have been successfully employed to simulate observed sequence distributions in polypropylenes, particularly with two state models(7,9,12,28,30). Results for the Doi 0/0 model typically predict a distribution where the asymmetric chain strongly dominates with only a low amount of a fairly syndiotactic symmetric chain component. In this study, the original Doi 0/0 two-state model was applied to NMR sequence distribution data from the cold *xylene insoluble* fraction of a highly isotactic polypropylene homopolymer taken from the first stage of an ICP sequential polymerization.

Results after fitting the observed methyl pentad/heptad distribution with the Doi 0/0 2-state model are given in Table 3. As observed by others(16,17), the fit nicely

Table 3. Observed Methyl Pentad/Heptad Distributions and Statistical Fits using Doi 2-State 0/0 Model for the Homopolymer Component of an ICP prepared with a DCPMS Donor.

Pentad/heptad	PP homopolymer component from ICP	
	Obs'd	Cal'd from Doi 0/0
mmmm	0.9862	0.9863
mmmr	0.0042	0.0041
rmmr	0.0016	0.0004
mmrr	0.0036	0.0039
mmrm+rmrr	0.0010	0.0016
rmrm	0.0006	0.0008
mrrrm	0.0001	0.0001
mrrrr	0.0001	0.0001
rrrrr	0.0003	0.0001
rmrrm	0.0000	0.0002
mmrrrm+rrrrmr	0.0001	0.0003
mmrrr	0.0003	0.0002
rmrrmr	0.0003	0.0001
mmrrmr	0.0002	0.0002
mmrrmm	0.0014	0.0017
% asym. Chain (ω)		99.4
std dev.		0.0004

reproduces the profile of the observed pentad/heptad distribution. The standard deviation between calculated and observed pentad/heptad fractions is also highly satisfactory. Two observations that can cause concern with the calculated result are the surprisingly high percentage of asymmetric chains (ω) and the low fraction obtained for [*rrrrr*]. The statistical parameters for the fit are given in Table 4. It

Table 4. Statistical Parameters for the Doi 2-state Model for ICP Polypropylene Homopolymer Component.

Sample	P_0	P_m	ω (% asym. Chain)
PP from ICP	0.9984	0.5377	99.4

also should be observed that the P_m parameter from the best Doi 0/0 fit indicates that the symmetric chain component is virtually atactic. It would be surprising if a separate symmetric chain component with a P_m around 0.5 would be found in the xylene insoluble fraction. The original Doi 0/0 model was re-examined in a former series of polypropylenes(18) where the DCPMS donor concentration was incrementally changed from 0 to 1.00 mmol. Xylene solubility information and isotactic levels are given in Table 5. Carbon 13 NMR spectra are shown in Figure 1.

Table 5. **Experimental Data for Isotactic Polypropylenes Made with a Fourth Generation Supported Ziegler Natta Catalyst**

Sample	DCPMS (mmol)	H2 (psig)	% Cold Xylene Sol.	% Cold Xylene Insol.	[*meso*] (Xyl. Insol.)	Avg. MRL (Xyl. Insol.)
PP-A	0.0	4	33.08	66.94	0.908	40
PP-B	0.0	80	27.93	72.12	0.941	50
PP-C	0.006	80	1.85	98.01	0.980	170
PP-D	0.050	80	0.99	98.94	0.990	290
PP-E	0.400	80	0.67	99.37	0.991	360
PP-F	1.000	80	0.56	99.46	0.991	430

The behavior of the symmetric chain component, predicted by the Doi 0/0 model, as the overall stereoregularity increases is shown in Figure 2. The Doi 0/0 fit indicates that the symmetric chain is essentially syndiotactic for polypropylenes prepared without a donor. When DCPMS is added to the catalyst system, the predicted symmetric chain component becomes atactic as can be seen in Figure 2. The Doi 0/0 model was applied to the cold xylene insoluble fractions throughout this series, which raises the question of whether such polymers would be insoluble in cold xylene. A second experimental observation that is inconsistent with the Doi 0/0 results is that the ^{13}C NMR data show the same resonance patterns consistently from PP-A through PP-F. As seen in Figure 1, it is primarily the intensities of the total resonance pattern and not the development of a new pattern that is observed experimentally. It can be clearly seen in Figure 2 that a new pattern is predicted for the symmetric chain component after the first addition of donor (PP-C). The question to be addressed at this point is whether a revised version of the Doi 2-state model can meet the above concerns and predict that the observed syndiotactic sequences connect isotactic blocks(17,18), as opposed to existing as a separate component within the composition distribution.

In an earlier study(18), it was shown that first order symmetric chain Markovian statistics predict syndiotactic chain defects even for very highly isotactic polypropylenes. These observations bring up the question of whether the solution for the Doi 2-state model, given in Tables 3 and 4 is unique. It should be noted that ω is a function of P_m. In the Doi 2-state pentad/heptad descriptions, each pentad/heptad fraction has the form $(1-\omega)$ times some power of P_m and/or $(1-P_m)(12)$. It needs to be established whether lower values of ω, when employed with correspondingly higher values of P_m, could also lead to satisfactory overall fits of the Doi 2-state model for pentad/heptad distributions.

Before proceeding further with any modified homopolymer statistical analyses, it is interesting to examine the copolymer compositional distributions for ICP-A and ICP-B in terms of the relative amounts of a-E/P versus c-E/P. The homopolymer component discussed in the preceding section was extracted from ICP-A. ICP-B is an

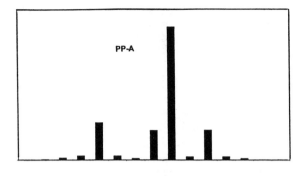

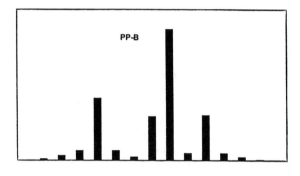

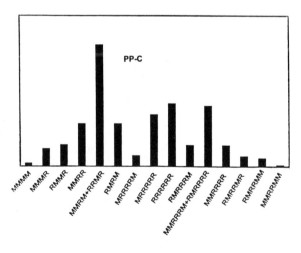

PENTAD/HEPTAD

Figure 2. The Symmetric Chain Pentad/Heptad Distribution Predicted by the Doi 0/0 Two State Model for Polypropylenes with Increasing Isotactic Levels(18)

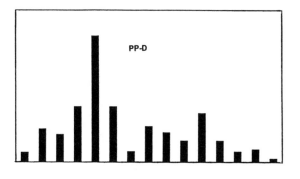

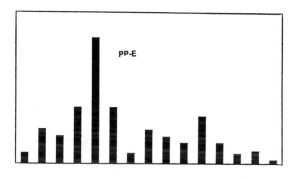

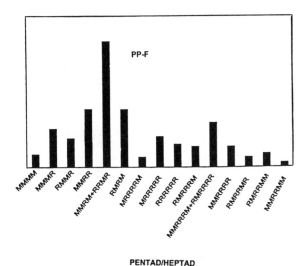

PENTAD/HEPTAD

Figure 2   Continued

experimental duplicate of ICP-A. The amount of amorphous E/P, relative to the total E/P, was 81 weight per cent for ICP-A and 85 weight per cent for ICP-B, as shown in Table 6.

Table 6. Distribution between a-E/P and c-E/P in ICP-A and ICP-B

Sample	% a-E/P (in copolymer)	%E (in total copolymer)	r_1r_2 for a-E/P	r_1r_2 for c-E/P
ICP-A	81.1	49	0.89	4.3
ICP-B	84.8	46	0.93	9.1

The differences between apparent r_1r_2s between amorphous and crystalline E/P copolymers have been used in an earlier study to indicate that these E/P copolymers originate from different catalyst sites(20). If that is the case, then the catalyst site distribution for ICP-A and ICP-B ranges from 15 to 19 % for the site producing crystalline E/P copolymers and 81 - 85 % for the sites producing amorphous E/P copolymers. Such a gravimetric procedure for determining the catalyst site distribution, of course, involves the implicit assumption that the activities of the two types of catalyst sites are similar. This may not be true as will be discussed later. The ratio of asymmetric/symmetric chains for the homopolymer predicted by the Doi 2-state 0/0 model is 99/1. This ratio of the two different types of homopolymer components chains differs substantially from the observed 85/15 amorphous/crystalline E/P copolymer distribution. One cannot necessarily stipulate that the ratio of catalyst sites that produce different types of homopolymers should necessarily be the same as those that produce different types of copolymers, but it is interesting to begin with that supposition.

There are some similarities in behavior between the first stage polypropylenes and the second stage E/P copolymers. For example, the Doi 2-state model, which utilizes Bernoullian asymmetric chain statistics, must by definition have r_1r_2s of unity for the predicted asymmetric chain sequence distributions. A test utilizing first order Markovian statistics for the polypropylene asymmetric chain component in the Doi 2-state model resulted in P_{00}s and P_{10}s so close that the use of first order Markovian statistics in place of Bernoullian statistics was unwarranted(18). The Bernoullian asymmetric chain behavior corresponds to the behavior of the more abundant a-E/P components, which have observed apparent r_1r_2s close to unity(20). The crystalline E/P components give considerably higher values for r_1r_2, which has been interpreted as a consequence of a broad compositional distribution(20.). This result corresponds to the first order Markovian symmetric chain behavior, which may also be consistent with a broad composition distribution(18).

In an earlier stage of this study(18), a Doi 2-state model was found to give good polypropylene homopolymer fits when tested with a Bernoullian asymmetric chain component and a first order Markovian symmetric chain component. This 2-state model is referred to as Doi 0/1. Several minima could be located with the Excel solver routine when fitting the observed pentad/heptad distribution with the Doi 0/1 parameters. The very best overall fit was close to that obtained with Doi 0/0(18). Other satisfactory fits with the desired lower ω can also be obtained and the pertinent parameters are given in Table 7. It is clear from the results in Table 7 that the Doi 0/1 solutions are not necessarily unique. In this case, satisfactory fits, within experimental error, were obtained with ω ranging from 0.8 to 0.99. The importance of using first order Markovian statistics for the symmetric chain component is that it

Table 7. Polypropylene Statistical Parameters from the Doi 0/1 Model

Sample	P_0	P_{mm}	P_{rm}	ω	Std. Dev.
PP from ICP	0.9985	0.5714	0.4894	0.993	0.0004_1
	0.9986	0.9902	0.4898	0.85	0.0006
	0.9988	0.9925	0.3366	0.80	0.0005_5

allows the fits with lower values of ω to be obtained. One of the consequences of the high ω result (0.993) in Table 7 is that the predicted symmetric chain component is approaching Bernoullian behavior. The lower values of ω also work well in Doi 0/1 as shown by a comparison of the calculated versus observed pentad/heptad distribution in Table 8 for $\omega = 0.80$.

Table 8. Observed Methyl Pentad/Heptad Distribution versus Statistical Fit Using Doi 0/1 2-State Model and $\omega = 0.8$

Pentad/heptad	PP Homopolymer component from ICP-A	
	Obs'd	Cal'd from Doi 0/1
mmmm	0.9862	0.9865
mmmr	0.0042	0.0048
rmmr	0.0016	0.0001
mmrr	0.0036	0.0038
mmrm+rmrr	0.0010	0.0010
rmrm	0.0006	0.0000
mrrrm	0.0001	0.0002
mrrrr	0.0001	0.0006
rrrrr	0.0003	0.0006
rmrrm	0.0000	0.0000
mmrrrm+rrrrmr	0.0001	0.0004
mmrrrr	0.0003	0.0009
rmrrmr	0.0003	0.0000
mmrrmr	0.0002	0.0000
mmrrmm	0.0014	0.0013
% asym. Chain (ω)		0.80
std dev.		0.0005_5

It is important at this point to return to polypropylenes PP-A through PP-F(18), where DCPMS was added incrementally to the catalyst system, to see how these systems behave with $\omega = 0.8$. The Doi 0/1 best fits for PP-A through PP-F with ω set at 0.80, are shown in Figure 3. The behavior of the Doi 0/1 parameters, which is considerably different from the Doi 0/0 behavior in Table 4 and Figure 2, is shown in Table 9. The symmetric chain parameter, P_{mm} steadily increased with increasing donor concentrations. As shown in Figure 3, syndiotactic defects connecting isotactic blocks were predicted for all cases. The syndiotactic defects steadily decreased in concentration as the donor, DCPMS, was increased as was the case observed experimentally with ^{13}C NMR(18). One of the important results from the fits with ω set at 0.80 is that P_{mm} is always less than P_0. This observation leads to the prediction that asymmetric chains are more isotactic than symmetric chains, consistent with the fractionation results of Paukerri, et al.(31).

144

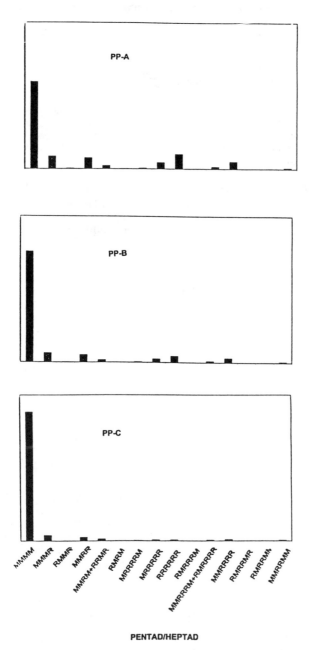

Figure 3 The Symmetric Chain Pentad/Heptad Distribution Predicted by the Doi 0/1 Two State Model for Polypropylenes with Increasing Isotactic Levels(18)

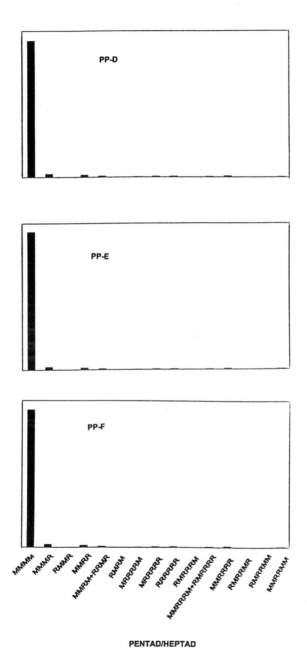

PENTAD/HEPTAD

Figure 3   Continued

There have been precedents for examining higher order Markovian statistical models for polypropylene. Second order Markovian statistics for symmetric chains has been suggested previously by Collette, et al.(32), but was applied to whole polymers and ether soluble fractions with little success. Later, R. Chûjo applied first order Markovian statistics to asymmetric chains for propylene polymers made with supported fourth generation Ziegler-Natta catalysts to show a preference for "00" over "11" diads with the addition of electron donors(33). One of the advantages of utilizing higher order Markovian statistics is that the systems will naturally reduce to lower orders if that should be the case. The asymmetric chain model did reduce to Bernoullian statistics, which was the reason for employing the Doi 0/1 two state model.

In Table 9, it can be clearly seen that both P_0 and P_{mm} increase systematically as the isotactic stereoregularity increases. The quality of the fits was also satisfactory as indicated by the standard deviations in Table 9. With ω fixed at 0.8, a completely different behavior is observed for P_{mm}, when compared to the original Doi 0/1 and Doi 0/0 fits where ω was allowed to find its optimum value during iterations with the Excel solver routine. P_{mm} increases with increasing donor concentrations and the predicted symmetric chain component is expected to be crystalline and insoluble in cold xylene. This result intuitively appears more reasonable than the original result where the symmetric chain component ranged from syndiotactic to atactic with increasing donor concentrations. Essentially no differences were observed for the fits with $\omega = 0.8$ and 0.85. Examinations of fits with values of ω down to 0.6 indicate that the Doi 2-state model is fairly insensitive to choices of ω in the 0.6 to 0.9 range. As stated previously, this behavior is caused by the fact that the calculated sequence

Table 9. Statistical Parameters for Solutions of the Doi 0/1 Model for a Series of Polypropylenes with Incremental DCPMS Additions from 0.0 to 1.0 mmol.

Sample	[*mmmm*]	P_0	P_{mm}	P_{rm}	ω	Std. Dev.
PP-A	0.8215	0.9743	0.9342	0.1794	0.80	0.0019
PP-B	0.8759	0.9813	0.9619	0.2205	0.80	0.0015
PP-C	0.9579	0.9942	0.9789	0.3524	0.80	0.0007
PP-D	0.9775	0.9986	0.9886	0.3208	0.80	0.0007
PP-E	0.9809	0.9992	0.9901	0.3396	0.80	0.0007
PP-F	0.9819	1.0000	0.9892	0.3239	0.80	0.0008

distributions involve variations of a basic function, $(1-\omega)$ times P_{mm}. It is clear from this study that the Doi 2-state model will not necessarily give unique values for ω.

Discussion of Results

The Doi 0/1 2 state model gives satisfactory fits over a broad range of ωs for the stereochemical pentad/heptad distribution observed for a highly isotactic polypropylene taken from the first stage of an ICP process. The ratio of crystalline to amorphous E/P copolymers prepared subsequently in the sequential polymerization process suggests that ω is nominally around 0.8. It is difficult to establish ω precisely from the E/P copolymer distribution because the relative activities of the sites producing amorphous versus crystalline copolymers are unknown. If the sites producing amorphous copolymers are more active than the sites producing crystalline copolymers, then ω may actually be lower than 0.8.

The fact that a broad compositional distribution of homopolymers exists even among highly isotactic polypropylenes has been observed using techniques other than the behavior of the sequence distributions. In Figure 4, a DSC scan is shown of the initial melting behavior of highly isotactic polypropylene granules taken directly from a reactor. This polypropylene has an average *meso* run length of about 400 and was prepared with the same catalyst system and DCPMS donor concentration as the polypropylene examined throughout this study. It is important to inspect the initial melting behavior of the polymer granules by DSC before the molecules have a chance to reorganize during an extrusion/pelletization process. After pelletization, the initial and subsequent DSC endotherms show only a single, broad peak. In Figure 4, four melting endotherms, ranging from 147 to 167 ° C are observed. This substantially broad range of stereoregularity has also been indicated by TREF studies(34). A distribution of more isotactic asymmetric chains versus lesser isotactic symmetric chains offers a reasonable and viable explanation for this observed DSC behavior. As suggested previously(18), the observed atactic polypropylene could arise from the low end of the symmetric chain compositional distribution.

The application of first order Markovian statistics to an analysis of sequence distributions takes into consideration the identity of the previous unit that added just before the unit of interest. This is not an unreasonable circumstance to consider when analyzing sequence distributions(35). Certainly, first order Markovian statistics should be used in a first attempt to fit any sequence distribution. If the observed distribution is truly Bernoullian or zero order Markov, then $P_{11} = P_{01}$ and $P_{00} = P_{10}$ and the system automatically reduces to zero order. This was precisely the case for the asymmetric chains of this study. If the catalyst site exerts enantiomorphic site control, then configuration is established solely by the catalyst site and zero order Markovian behavior should be expected.

Symmetric chains have essentially equal configurational populations and there may or may not be enantiomorphic site control. In the simplest case, $P_{00} = P_{11} = P_m$ and $P_{01} = P_{10} = P_r$. First order Markovian statistics are appropriate when $P_{mr} \neq P_{rr}$ and $P_{rm} \neq P_{mm}$. In this study, the symmetric chains did not reduce to zero order at high isotactic levels. It would be interesting to determine if first order Markovian statistics, where $P_{01/0} = P_{10/1}$ would innately include alternation between two enantiomorphic sites. A comparison of predicted syndiotactic sequence distributions from the alternating mechanism versus classical first order Markovian statistics will be the subject of an upcoming report.

One of the interesting results that arise from an application of first order Markovian statistics to polypropylene pentad/heptad sequence distributions is that different families of stereo defects are predicted for asymmetric versus symmetric chains(18). The observed methyl pentad/heptad distribution is a composite of independent resonance patterns arising from different types of stereo defects. Certainly a great deal of structural information is available after deconvolution of the ^{13}C NMR methyl resonance pattern. Metallocene catalysts are providing valuable reference polypropylenes for characterization purposes because metallocene catalysts can be chosen to produce correctly either symmetric or asymmetric chains and not a combination of the two.

Finally, we should be reminded of the statement by F. P. Price(6), which was quoted in the introduction. Markovian statistics are applied to analyses of polymer sequence distributions observed after polymerization is complete. The Markovian transition probabilities are between initial and final states and say nothing about the pathways leading to initial and final states. Consequently, considerable care should be invoked before using Markovian statistical behavior of sequence distributions to establish catalyst mechanistic behavior.

148

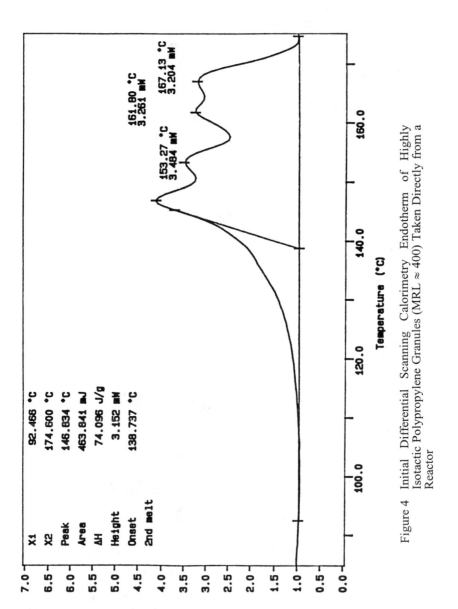

Figure 4 Initial Differential Scanning Calorimetry Endotherm of Highly
Isotactic Polypropylene Granules (MRL ≈ 400) Taken Directly from a
Reactor

References

1. Boor, J. Jr.; *"Ziegler-Natta Catalysts and Polymerizations,"* Academic Press, New York, 1979, Chapter 15
2. Wolfgruber, C.; Zannoni, G.; Rigamonti, E.; Zambelli, A.; *Makromol. Chem.* **1975**, 176, 2765.
3. Paukkeri, R.; Vaananen, T.; Lehtinen, A.; *Polymer*, **1993**, 34, 2488
4. van der Van, Ser; *Polypropylene and Other Polyolefins*; Elsevier Science Publishers B. V., Amsterdam, 1990, 266-288
5. Paul, D. R.; Barlow, J. W.; Keskkula, H.; *Concise Encyclopedia of Polymer Science and Engineering*, J. I. Kroschwitz, Executive Editor, John Wiley & Sons, N.Y., 1990, p.830
6. Price, F. P.; *"Markov Chains and Monte Carlo Calculations in Polymer Science,"* Marcel Dekker, G. G. Lowry, Editor, New York, 1970, Chapter 7
7. Shelden, R. A.; Fueno, T.; Tsunetsugu, T.; Furukawa, J.; *J. Polym. Sci.,* **1965**, Part B 3, 23
8. Bovey, F. A.; Tiers, G. V. D.; *J. Polym. Sci.* **1960**, 44, 173
9. Bovey, F. A.; *"High Resolution NMR of Macromolecules,"* Academic Press, New York, 1972
10. Randall, J. C.; *"Polymer Sequence Determination: Carbon-13 NMR Method,"* Academic Press, New York, 1977, Chapter 4
11. Farina, M.; Di Silvestro, G.; Terragni, A.; *Macromol. Chem. Phys.* **1995**, 196, 353
12. Doi, Y.; *Makromol. Chem., Rapid Commun.* **1982**, 3, 635
13. Hayashi, T.; Inoue, Y.; Chujo, R.; Asakura, T.; *Polymer*, **1988**, 29, 138
14. Coleman, B. D.; Fox, T. G.; *J. Chem. Phys.* **1963**, 38, 1065
15. Cheng, H. N.; Kasehagen, L. J.; *Macromolecules* **1993**, 26, 4774
16. Chûjo, R.; Kogure, Y.; Vaananen, T.; *Polymer*, **1994**, 35, 339
17. Busico, V.; Cipullo, R.; Corradini, P.; De Biasio, R.; *Macromol. Chem. Phys.,* **1995**, 196, 491
18. Randall, J. C.; *Macromolecules* **1997**, 30, 803
19. van der Van, Ser; *Polypropylene and Other Polyolefins*; Elsevier Science Publishers B. V., Amsterdam, 1990, pp 290-293
20. Randall, J. C.; *J. Poly. Sci.(A), Poly. Chem.* **1998**, 36, 1527
21. Stehling, F. C.; Huff, T.; Speed, C. S.; Wissler, G.; *J. Appl. Polym. Sci.*, **1981**, 26, 2693
22. Cozewith, C. *Macromolecules* **1987**, 20, 1237
23. Chadwick, J. C.; van Kessel, G. M. M.; Sudmeijer, O.; *Macromol. Chem. Phys.,* **1995**, 196, 1431
24. Chadwick, J. C. *"Advances in Polyolefins"*, Sept. 28 - Oct. 1, 1997, Napa, California
25. Kaye, H.; Polyhedron Laboratories; Private Communication
26. Randall, J. C.; *J. Polym. Sci., Polym. Phys. Ed.*, **1976**, 14, 1693
27. Schilling, F. C.; Tonelli, A. E.; *Macromolecules* **1980**, 13, 270
28. Busico, V.; Corradini, P.; De Biasio, R.; Landriani, L.;Segre, A. L.; *Macromolecules* **1994**, 27, 4521
29. Person, R.; *"Using Excel-4 for Windows"*, Que Corporation, Carmel, IN, 1992
30. Paukkeri, R.; Iiskola, E.; Lehtinen, A.: Salminen, H.; *Polymer* **1994**, 35, 2636
31. Paukkeri, R.; Lehtinen, A.; *Polymer*, **1994**, 35, 1673
32. Collette, J. W.; Ovenall, D. W.; Buck, W. H.; Ferguson, R. C.; *Macromolecules*, **1989**, 22, 3858

33. Chûjo, R.; *Stepol '94 Abstracts*, "International Symposium on Synthetic, Structural and Industrial Aspects of Stereospecific Polymerization", Milan, Italy, June 6-10, 1994, p. II-13
34. Ushioda T.; *SPO 97*, 101, Sept. 24-26, 1997, Schotland Business Research, Inc., Houston, TX 1997
35. Price, F. P.; *J. Chem. Phys.*, **1962**, 36, 209

Chapter 10

Short Chain Branching Distribution of Amorphous Ethylene Copolymers

Lecon Woo, Michael T. K. Ling, Atul R. Khare, and Stanley P. Westphal[1]

Corporate Research and Technical Services, Baxter International, Round Lake, IL 60073–0490

A new method is proposed to characterize the short chain branching distribution (SCBD) for amorphous ethylene copolymers. SCBD controls many important properties for polyethylenes. It was well known, for example, that at equal density or crystallinity, the optical clarity measured by % haze with different short chain distributions could be different by more than an order of magnitude. The recently introduced metallocene catalysts are capable of producing very homogeneous copolymers at densities heretofore near impossible. These developments re-emphasized the need for rapid assessments of SCBD. Dynamic mechanical analysis was carried out on a series of low crystallinity ethylene copolymers including very low density and ultralow density polyethylenes (VLDPE and ULDPE) and ethylene vinyl acetate (EVA). The beta relaxation linewidth when normalized for mole fraction of the comonomer, was found to correlate with the short chain branching distribution obtained by temperature rising elution fractionation (TREF). The proposed method for assessing SCBD is applicable for totally amorphous systems where TREF analysis loses its fractionation ability. Results on an ethylene propylene rubber system will be used to illustrate this point.

In the polyolefin industry, there existed a long-standing desire to create progressively lower density materials by copolymerization. These lower density olefins potentially offer a range of attractive properties like lower moduli (higher flexibility), greater toughness and lower heat seal initiation temperatures. Until recently, the lowest density commercially achievable remained near the .915 range. Any attempt to lower the density further invariably resulted in too much of an amorphous and low molecular weight fractions which rendered the polymer very difficult to handle. However, in the last few years new catalysts, most notably the metallocene compounds based on zirconium, coupled with improvements in process

[1]Current address: 3150 Excelsior Boulevard, Apartment 101, Minneapolis, MN 55416.

technology, such as series reactor designs, have resulted in a large number new products of ever-decreasing densities.

In the medical industry, these lower density olefins present very interesting possibilities for product designs. For example, the extremely high catalyst activity may lead to materials with very low inorganic contaminations, while the high comonomer content and statistical distribution may provide superior optical clarity for particulate matter (PM) inspection. In addition, some of the softer grades of materials possess good thermoplastic elastomeric properties for innovations in product design (1).

For polymers of equal comonomer content, very different properties can result depending on the way the comonomer is distributed on the main chain. The resulting short chain branching distribution (SCBD) is needed to fully characterize the polymer. The traditional Ziegler-Natta catalysts, due to their multi-site nature, produce very broad SCBD. As a result, the product consists of a mixture of highly branched (high comonomer content) low molecular weight species and less branched high molecular weight fractions. The preferred method to quantify this intermolecular branching distribution is temperature rising elution fractionation (TREF) (2). This solution technique physically separates polymer fractions based on differences in crystallizability at different temperatures. However, as polymer crystallinity decreases, the basis for the TREF separation is greatly diminished, and few other techniques are available.

The purpose of this work was to characterize the short chain branching distribution for highly amorphous (low crystallinity) polymers by thermal and rheological techniques. Materials used in this study were ethylene copolymers with butene, octene and vinyl acetate. Polymer types covered include linear low density polyethylene (LLDPE), very low density polyethylene (VLDPE), ultra low density polyethylene (ULDPE), ethylene vinyl-acetate (EVA) and ethylene-propylene rubber (EPR). For our purposes, the LLDPE covers the density range between 910 to 925 Kg/m^3, VLDPE's between 890 to 910 and ULDPE 's with densities of 890 and below.

EXPERIMENTAL:

Ethylene copolymer samples examined in this study included: ULDPE #1, a solution polymerized butene copolymer with specific gravity of 883Kg/m^3 and melt index of 3.6 dg/min, ULDPE #2, a high pressure process butene copolymer of 880 density and 2.2 melt index, ULDPE #3, a gas phase butene copolymer with 890 density and 1.0 melt index, ULDPE #4, an octene copolymer of 870 density and melt index of 2.0, a solution process octene VLDPE of 912 density and 3.3 melt index, EVA#1 is a 18 wt % copolymer with a 0.45 melt index from the high pressure process, EVA# 2 is a 28wt. % copolymer with a 3.2 melt index, and ethylene propylene rubbers (EPR) with 45 and 75wt. % ethylene contents respectively for the

#1 and #2 samples. Pellets of the samples were first compression molded into ca .2 mm thick films on a compression molding press at 190° C. Then their thermal properties measured on a TA Instruments 2910 Differential Scanning Calorimeter (DSC) cell using a 2100 controller. Heating and cooling rates of 10°C and 5°C min-1 were used throughout this study. A Seiko Dynamic mechanical Analyzer (DMA) DMS-110, was used for wide frequency range (0.5- 100 Hz) studies over the temperature range of -150 to 150°C. For DMA studies , compression molded samples of about 1 mm in thickness were used in the three point flexural mode. The temperature program rate was at 3°C min-1. Although multiple frequency data can be used for activation energy calculations, for the purpose of this work, data at the reference frequency of 1 Hz was used. Morphological studies were also carried out using a Reichert FC4E cryo-ultramicrotome to prepare undistorted material blocks for SEM analysis. SEM imaging was done with the JEOL 6300 FESEM after sonic bath etching in ambient temperature heptane. In addition, other available characterization data were incorporated into this study.

RESULTS AND DISCUSSION:

DSC (Differential Scanning Calorimetry) analysis of polyethylenes characterizes the crystalline character of the material. Selected second heat DSC data are shown in Table I.

TABLE I
THERMAL ANALYSIS OF POLYETHYLENES

SAMPLE	Tm(C)	ΔH_f (J/g)	% CRYSTALLINITY
ULDPE#1	69.5	59.8	20.4
ULDPE#2	71.8	62.0	21.2
ULDPE#3	118.4	74.8	25.6
ULDPE#4	55.3	38.0	13.0
VLDPE	123.8	125.6	42.9
EVA#1	84.1	74.8	25.6
EVA#2	69.9	60.3	20.6
EPR #1	(20)	-- --	ca.0.0
EPR #2	47.9	35.9	12.3

In Table I, ΔH_f = 292 J/g is taken for 100% crystalline polyethylene.

Among the samples examined, there was a wide range of melting temperatures and crystallinities. These values only reflect part of the material response. The melting curve is a reflection of the lamella thickness distribution, which is in turn a function of sample history and composition. In several of these materials the melting curve exhibits multiple endothermic peaks even though the material is a single grade of polyethylene. Multiple melting peaks indicate several distinct lamella thickness populations. This indicates that the comonomer distribution is heterogeneous. This type of melting response occurs with most LLDPE type resins due to the multiple catalytic active sites with different reactivities toward comonomers. Much narrower melting curves are seen in conventional linear HDPE homopolymers, high pressure LDPE or EVA copolymers.

Although we can detect heterogeneity among crystallizable components with DSC, it is difficult to quantify because of the difference in specific heat accompanying increased comonomer incorporation. Another area of difficulty occurs when one or more of the components is amorphous. The observed effect will be to reduce crystallinity at the same melting temperature. This is illustrated in Figure 1. Two samples, ULDPE#3 and the VLDPE, exhibit a similar melting temperature but very different crystallinities.

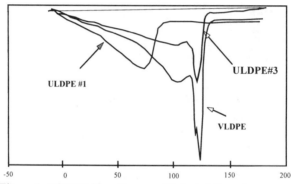

Figure 1, DSC Melting Curves of Selected Samples

Figure 2 is a heptane etched surface morphology of ULDPE #3 by scanning electron microscopy, with the white scale bar 2 micron in length. For this low crystallinity sample, it is evident through its very heterogeneous structural distribution, high melting and etch resistant domains existed in a continuous network. Due also to this heterogeneity, the optical clarity is rather poor. A 46% haze was measured by the ASTM-1003 method for a 125 micron film, compared with about 4% haze for more homogeneous samples like ULDPE #1.

The most effective way to characterize polyethylene heterogeneity is through temperature rising elution fractionation (TREF). This technique fractionates material based on differences in crystallizability during very slow crystallization from dilute solutions. Application of this technique to various polyethylenes is illustrated in Figure 3. It is very obvious that the short chain branching distribution is very different for the three samples. This short chain branching distribution can be readily quantified in a manner analogous to size exclusion chromatography (SEC). This is illustrated below:

$$SCBD = \frac{(\Sigma \, Ni \, SBi^{2}) \, / \, (\Sigma \, Ni \, SBi)}{(\Sigma \, Ni \, SBi) \, / \, (\Sigma \, Ni)} \qquad (1)$$

where: Ni is the weight (detector response) of given fraction
SBi is the short chain branching (CH_3/1000 carbons)

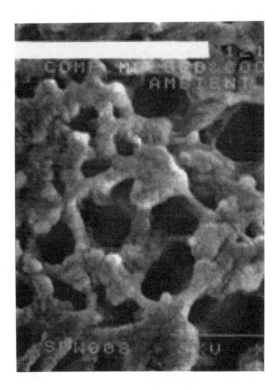

Figure 2, SEM morphology of heptane etched ULDPE #3

TREF is the best technique for quantifying polyethylene SCBD provided that the materials are mostly crystallizable and the heterogeneity is intermolecular. However, it suffers from the same difficulty as DSC analysis; amorphous fractions can not be

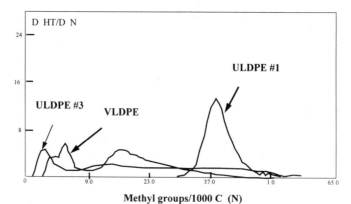

Methyl groups/1000 C (N)

Figure 3, TREF Analysis of Selected Samples.

fractionated. Typically, as the comonomer content approaches about 20 mole percent, the residual crystallinity becomes negligible. Thus the distribution of branches in the amorphous phase can not be fractionated based on crystallizability. In addition, the technique is also very time consuming and labor intensive without automation.

Dynamic Mechanical Analysis (DMA) is sensitive to the structural heterogeneity of polyethylene (3). During the dynamic mechanical analysis, the sample is subjected to a periodic sinusoidal deformation, and the resulting stress monitored and deconvoluted, into "in-phase" and "out of phase" components: elastic and loss moduli (Figure 4). Their ratio is defined by the tangent of the phase angle δ. As a sample is programmed over a temperature range, typically between the deep cryogenic temperatures of about -150 °C to the softening or melting points, a series of relaxation events can be recorded as a step reduction in the elastic modulus, accompanied by a relaxation maximum in E" or tan δ.

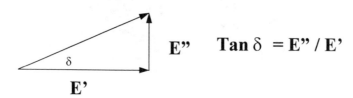

Figure 4, Dynamic Mechanical Analysis

The beta relaxation has been shown to originate from the amorphous phase near the branching points (4) . The beta relaxation's location and intensity depend on structure and density of branching. For ethylene vinyl acetate copolymers (EVA), the beta transition temperature decreases steadily with increasing vinyl acetate content to a minimum before turning upward for vinyl acetate rich copolymers (Figure 5). The functional behavior for a minimum in beta relaxation temperatures has been observed for a variety of copolymers with ethylene. For example, in the ethylene propylene system, after the minimum is reached, the beta relaxation returns toward 0°C, the glass transition for amorphous polypropylene. For lower content (below 40 wt.%) vinyl acetate copolymers, since the melting temperature is a function of VA content, the melting temperature also correlates with the beta relaxation temperature. In general, the beta relaxation temperature correlates with observable melting temperature in other ethylene copolymers. The steady decreasing beta relaxation for increasing branch content copolymers is the basis we would like to exploit for branching distribution determination.

It has been noted previously that the width of the beta relaxation qualitatively correlated with the short chain branching distribution (SCBD) defined from the TREF analysis (5). Further, the linewidth is also influenced by the comonomer content. EVA copolymers polymerized at high pressure via free radical process are known to exhibit a rather narrow comonomer distribution (2).

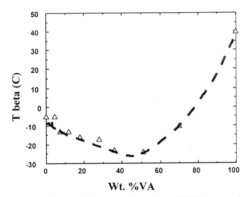

Figure 5, Beta relaxation of EVA copolymers.

In DMA, a measure of the breadth of the relaxation process is the full width at half height (FWHH). Early data by Schmieder and Wolf (6) on chlorinated polyethylenes (another random process) fit nearly identically as EVA on a molar basis (Figure 6). This identical dependence on very different branch units indicates that at least for ethylene copolymers, this phenomenon is quite universal. Hence the linewidth dependence of EVA and chlorinated PE can be used as a reference point representing a narrow SCBD case.

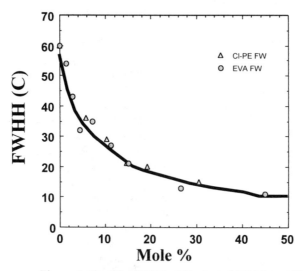

Figure 6, Combined EVA, Chlorinated PE Linewidths.

For the time being, the slight complications of long branches in these reference cases will be ignored, leaving open the possibility that some narrowly distributed linear copolymers could exceed these reference cases in narrowness in SCBD. Three samples with wide compositional distributions which were analyzed in detail with both DSC and TREF were chosen for dynamic mechanical analysis (Figures 7-9). It was seen that they varied greatly in beta linewidth. For the determination of beta relaxation linewidths, a baseline was drawn from the high temperature side of the gamma relaxation at about -120°C parallel to the temperature axis. Full width half height (FWHH) in °C was used for the measurement of relaxation linewidth.

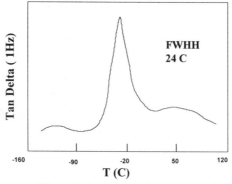

Figure 7, Tan Delta of ULDPE #1.

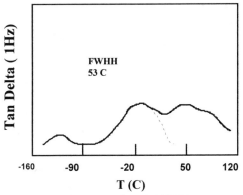

Figure 8, Tan Delta of VLDPE.

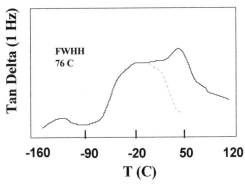

Figure 9, Tan Delta of ULDPE #3

After normalizing the beta relaxation FWHH to that of the EVA or chlorinated polyethylene at equal molar comonomer contents, we find an excellent dependence with the SCBD determined by TREF (Figure 10). This normalized linewidth SCBD plot can then be used to determine distributions of unknown copolymers. Furthermore, the same "calibration" plot can be extended for non-crystalline samples beyond the fractionation capabilities of TREF.

The proposed procedure is then as follows: 1), determine the molar comonomer content based on density or crystallinity, 2), measure the beta relaxation linewidth at 1 Hz from the dynamic mechanical spectrum, 3), calculate the normalized linewidth from Figure 6, and 4), determine the SCBD from Figure 10.

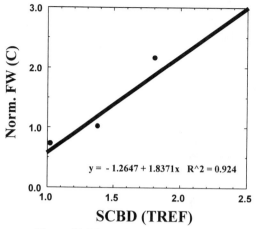

Figure 10, Normalized Linewidth vs. SCBD.

To test the validity of this hypothesis, we applied the method to two ethylene-propylene rubbers (EPR) containing 45 wt.% ethylene and 75 wt.% ethylene which exhibited 0% and 12 % crystallinity, respectively (Figures 11-12). The normalized beta linewidth of the amorphous EP rubber yielded a SCBD <1.01, a very narrow branching distribution. The normalized beta linewidth of the second product yielded a SCBD of 2.40, a much broader distribution. These two materials were both produced by the same supplier. We suspect that the two materials were made with different catalyst systems and process conditions resulting in very different SCBD. Reportedly, for the ethylene propylene system, vanadium based catalysts produce very narrow distributions while titanium catalyzed products very broad.

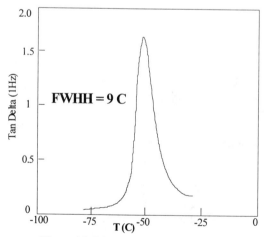

Figure 11, DMA Tan δ of EPR #1

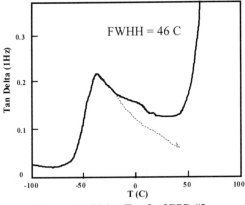

Figure 12, DMA Tan δ of EPR #2

The DMA technique works very well with amorphous to low crystallinity materials like VLDPE and ULDPE. For higher crystallinity materials of 920 or greater density, the beta relaxation became indistinct and merged into part of the alpha relaxation (7). This overlapping relaxation and the accompanying reduction in beta intensity makes the technique less sensitive above these densities. For example, Figure 13 shows the DMA spectrum for a Ziegler Natta butene copolymer of 920 density linear low density polyethylene (LLDPE) from a gas phase process. The beta relaxation at about -20°C, although very broad, became submerged on the shoulder of the much more prominent α and α' relaxations. This reduction in intensity and the interference from the alpha relaxations renders the quantitative analysis difficult without extensive spectral deconvolution. Of course, at these higher densities, the traditional TREF techniques would still be the most suited for detailed quantization.

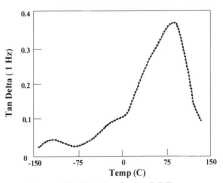

Figure 13, DMA of a LLDPE.

SUMMARY:

A new method for characterizing SCBD of highly amorphous ethylene copolymers was proposed based on dynamic mechanical analysis on whole polymers. Dynamic mechanical analysis was carried out on a series of low crystallinity ethylene copolymers including very low density and ultralow density polyethylenes (VLDPE and ULDPE) and ethylene vinyl acetate (EVA). The beta relaxation linewidth when normalized for mole fraction of the comonomer, was found to correlate with the short chain branching distribution (SCBD) determined by temperature rising elution fractionation (TREF). The proposed method of using normalized beta relaxation linewidth to determine polyethylenes SCBD was shown to be promising for the systems studied. The method is particularly amenable to amorphous polymers where traditional techniques like DSC and TREF fail. Since the DMA method works on whole polymers without extensive sample preparation, rapid results can be obtained in as short as 2 hours. If supported by further data, this method could become a useful and complementary technique for polymer analysis.

ACKNOWLEDGEMENT:

Permission to publish this work by Baxter Healthcare is gratefully acknowledged. Also we are indebted to Les Wild for arrangement of the TREF analysis from the former Quantum Chemical Laboratories.

REFERENCES:

1. L. Woo, S.P. Westphal, M.T.K. Ling, NATAS Conf. **1992**, 39.
2. L. Wild, T.R. Ryle, D.C. Knobeloch, I.R. Peat; J.Poly.Sci., Phys, **1982**.,20, 441.
3. R. Boyd, Polymer, **1984**, 26, 1123.
4. R. Popli and L. Mandelkern, Polymer Bulletin , **1983**, 9, 260.
5. L. Woo, M.T.K. Ling, S.P. Westphal, ACS-PMSE , **1993**, 69, 316.
6. K. Schmieder and K. Wolf, Kolloid Z., **1953**, 134, 149.
7. N.G. McCrum, B.E. Read, G. Williams, <u>Anelastic and DielectricEffects in Polymeric Solids</u>, **1967,** 366, J. Wiley, New York.

Chapter 11

Variable Temperature, Solid-State ^{13}C NMR Study of Linear Low-Density Polyethylene

Mingming Guo, Stephen Z. D. Cheng, and Roderic P. Quirk

Institute of Polymer Science, The University of Akron, Akron, OH 44325–3909

Several variable-temperature solid-state ^{13}C NMR methods have been carried out on copolymers of ethylene and 1-octene, 1-hexene, 1-butene, and vinyl acetate prepared using both Ziggler-Natta and metallocene catalysts. These methods are presented as an attractive approach for the determination of comonomer types and contents in polyolefins as well as cross-linked polyolefins. These procedures also provide information about the crystal structure, crystallinity, interfaces, and branch location of the samples. New results concerning the crystallization, molecular motion and morphology of polyolefins synthesized via metallocene catalysts have also been demonstrated. The advantages of the *in-situ*, solid-state VT NMR methods to rapidly extract the heterogeneity and molecular segregation kinetics information of these polyolefins are demonstrated.

INTRODUCTION

Linear low-density polyethylene (LLDPE) is a material obtained by the copolymerization of ethylene and α-olefins such as propene, 1-butene, 1-hexene, etc.[1] Depending on the catalytic system as well as the type and amount of the comonomer introduced into the polymeric chain, a wide range of product grades with different physical properties is obtained[2-4]. These linear lower density polyolefins potentially offer a range of attractive properties like higher flexibility, greater toughness and lower heat-seal initiation temperatures, compared with traditional linear high density polyethylene, and non-linear high density polyethylene. The most important parameters characterizing these copolymers are molecular mass, molecular mass distribution, amount and type of the 1-olefin (side chain length), and intra- and intermolecular chemical composition distributions. The known methods to study the side chain distribution of these LLDPEs are based on the coupling of different analytical techniques such as thermal rising elution fractionation (TREF) and IR or NMR analysis. Recently, an improved fractionation method has also been reported.[5]

Solution ^{13}C NMR is generally regarded as the only analytical technique capable of identifying and quantifying all of the branching features of a polyolefin.[6-8] However, it suffers from four notable problems.[8-9] The first is that the sample must be dissolved into an appropriate solvent. As a result, insoluble polyolefins and crosslinked systems are not amenable to this approach. The second problem is its relatively low sensitivity. Overnight data acquisition times are commonly required to obtain satisfactory quantitative results due to the concentration and relatively long pulse delay requirements. The third problem is that it cannot provide any information about the bulk structures, such as chain packing, morphology, crystallinity, location of the branch, etc. The last problem is that it cannot provide long-chain sequence distribution information, although gaga hertz NMR is available recently.

In this paper, we report on the use of variable temperature, solid-state NMR to determine comonomer types and compositions for linear and crosslinked polyolefins. In addition, the long sequence distribution, branch location, and molecular segregation kinetics of the samples are characterized, which demonstrate the usefulness of the extension of the VT MAS NMR technique to characterize polyolefins. We believe that these approaches have important implications to provide not only static information but also kinetic information for polyolefins.

EXPERIMENTAL SECTION

The samples chosen for this study are commercially available copolymers of ethylene and 1-hexene (EH), and 1-butene (EB), (Exxon Chemical Company) and vinyl acetate (Aldrich). Sample EH and EB were prepared using metallocenes as catalysts. The molecular weight of samples of EH and EB are 98 K and 10.6 K, the polydispersities (M_w/M_n) are 2.21 and 1.93, respectively. The average short chain branch contents of these two samples are 7.8 and 47.2 per 1000 total carbon atoms based on solution ^{13}C NMR measurements. There are 9 mol % vinyl acetate units in the EVA sample.

In order to study molecular motion and morphological behavior of the copolymers of ethylene and 1-hexene (EH), this sample was prepared under three different conditions. The step annealed sample EH-sa was isothermally crystallized at each temperature in a stepwise manner decreasing from the isotropic melt as described in ref. 10. In brief, each step was 5 °C descending from the previous annealing temperature. Total annealing time was 72 hours. For comparison, the single step annealed sample EH-a was isothermally crystallized at 85 °C for 72 hours from the isotropic melt. Sample EH-q was quenched from the molten state directly into dry ice-acetone bath at −78 °C.

The variable temperature, solid-state NMR experiments were performed on a Chemagnetics CMX 200 NMR spectrometer operating at 50.8 MHz using a

Chemagneties PENCIL probe with 7.5 mm zirconia rotors. Spinning speeds were about 4.5 kHz, sufficient to eliminate spinning side bands. Melt state direct polarization magic-angle spinning (DPMAS) ^{13}C NMR was conducted with 4.3 μs 90° pulses for ^{13}C and 58-kHz decoupling gated on during signal acquisition. The decoupler frequency (about 1.5 ppm from TMS in the ^{1}H frequency domain) and amplitude (58 kHz) were adjusted to give the narrowest possible line width for the orthorhombic peak at room temperature. Chemical shifts were referenced to TMS in solution via the aromatic carbon line of hexamethyl benzene (17.35 ppm) in the solid. The solid temperature was controlled using a microprocessor-based controller, FEX-F900, PKC Instrument Inc., Japan. For the crystallization kinetics studies, the probe was retuned each time after the temperature was changed.

Proton $T_{1\rho}$ relaxation times were measured as the slope of the intensity of the ^{13}C CP/MAS NMR spectra vs. the time of proton spin locking. The pulse sequence employed was ^{1}H $(90°_x)(90°_y, \tau)$ followed by simultaneously 1 ms ^{13}C and ^{1}H spin lock and then acquisition of the ^{13}C magnetization with proton dipolar decoupling. The length of τ ranged from 0.1 to 40 ms.. A total of 16 τ values were usually recorded to determine each slope.

The ^{13}C and ^{1}H wide-line spectra detected by two-dimensional wide line separation spectroscopy (WISE)[11] were acquired with 64 t1 increments of 5 μs with a 1 ms cross-polarization contact time. Incrementing the proton evolution time, t_1, leads to a modulated ^{13}C signal, and subsequent two-dimensional Fourier transformation yields a spectrum with dipolar-broadened proton lines resolved in ω, by the carbon MAS spectra.

RESULTS AND DISCUSSION

Fast Determination of Comonomer Types at Room Temperature

The microstructure of ethylene copolymers depends on the comonomer type, the comonomer content, the distribution of the comonomers and branches along the chains, and the molecular weight distribution of the copolymers. There are many current analytical techniques that are capable of addressing most of these subtitles. This and following sections focus on approaches to rapidly estimate comonomer types and contents for the most common commercial polyolefin copolymers. Figure 1 shows the ^{13}C CP/MAS NMR spectra of four LLDPE samples recorded at room temperature. Although the resolution is poor, the fast room temperature ^{13}C CP/MAS NMR spectra (100 scans with total acquisition time of about 8 min) of four polyolefins copolymers EO, EH, EB, and EVA show the small, recognizable and unique methyl resonance of the butane branch at 11 ppm, hexane or longer branches at about 15 ppm, and the vinyl acetate carbonyl as well as -CH-O at 74 ppm, respectively. This fast and routine method can be used to quickly examine unknown polyolefin copolymer resins. The combination of spectral broadening, conformation differences of the samples, and varying cross polarization effects renders this simple method to be ineffective in determining comonomer type and content, however.

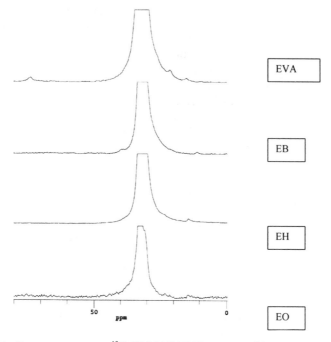

Figure 1. Room temperature ^{13}C CP/MAS NMR spectra of four LLDPE
samples. EVA: poly(ethylene-co-vinyl acetate); EB: copolymer of ethylene
and 1-butene; EH: copolymer of ethylene and 1-hexene; and EO: copolymer
of ethylene and 1-octene

Determination of Comonomer Types and Contents in the Melt State

Figure 2 shows the ^{13}C CP/MAS NMR spectra of four ethylene/α-olefin
copolymers EO, EH, EB and EVA in the melt state at 200 °C. Due to the dramatic
increase of the mobility of the sample at the melt state, these spectra have very nearly
the same level of resolution as those typically obtained in solution. The small
difference of the side chain resonances between EH and EO can be recognized. The
chemical shifts of Cα and C$_3$ are overlapped at 34.6 ppm in the EO case, while in the
EH case they are at 34.2 and 34.6 ppm, respectively. By integrating the NMR spectra
using a 30 s pulse delay, it was possible to calculate the comonomer content. The
branch amounts of the EH and EB samples are 6.8 and 41.2 per 1000 carbon atoms
based on the melt state NMR spectra, which are close to the values obtained using
solution NMR.

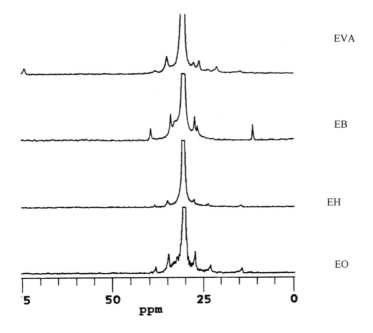

Figure 2. Melt-state (200 °C) ^{13}C MAS/DD NMR spectra of four ethylene/α-olefin copolymers EO, EH, EB and EVA

restricted to comonomer type and content determinations. It has limited utilities for analysis of cross-linked polyolefin frequently used in polyolefin industry mainly because they only swell in the presence of solvent. The problems of investigating these intractable crosslinked polyolefins may be overcome using solid state ^{13}C MAS, high-power ^{1}H decoupling NMR at elevated temperatures. In order to demonstrate that the melt state MAS NMR technique extends the ability of NMR, we have applied it to a crosslinked system. The sample was prepared by irradiating a pressed film of an 80/20 w/w EO and EVA blend. The spectra in Figure 3 show that the 2% vinyl acetate signal, resonating at 75 ppm in the cross-linked sample can be detected by using the melt state MAS NMR. The spectra of the same samples were recorded at ambient temperature (spectra not shown here). The broad lineshape of the sample at this condition prevented the detection of the small resonance of the PVA component.

Crystallinity Determination

The backbone methylene resonance of polyethylene shows separable signals at 33.3, 32.0 and 30.1 ppm arising from the crystalline and amorphous regions.[12-15]

The integral intensities of the lines do not quantitatively correspond to the crystallinity because of the difference in cross polarization dynamics as well as in proton relaxation for the different phases in the sample. Variable contact time experiments are frequently used to extract the crystallinity of polyethylene.[13] Before applying this technique, we need consider the proton T_1 effect. $^HT_1$ for most polymers

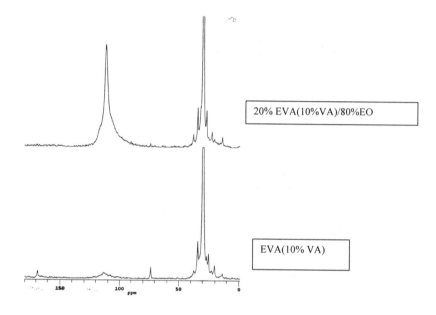

Figure 3. Melt state MAS NMR of EVA (10% VA) and the crosslinked polyolefins prepared by irradiating a pressed film of a 20% (10% VA) EVA and 80% EO blend. X is the signal from the Teflon sample tube spaces.

is about 1 s. A 3 to 5 s pulse delay time is frequently used to obtain ^{13}C CP/MAS spectra of polymers.[12-18] This value is suitable for the copolymers of ethylene and α-olefin. However, for some very rigid PE samples this value is not adequate. We measured the $^HT_1$ of the gel-spun, ultrahigh molecular weight PE fiber, Spectra 1000.[19] The $^HT_1$ of orthorhombic and monoclinic crystals are 2.6 s and 3.1 s respectively. Therefore, a 12 s pulse delay time was used to obtain the variable contact time CP/MAS spectra of the fiber sample. A solid echo ^{1}H NMR technique was also used to estimate the crystallinity of the SPECTRA 1000 sample.

Interface Region

In terms of the chemical shift, spin-lattice and spin-spin relaxation times of solid state high resolution ^{13}C NMR, it is possible to not only explicitly characterize the crystalline and amorphous phase in the semicrystalline polymers but also to determine the existence of an interface region as well.[20] Using the differences in ^{13}C spin-lattice relaxation time (T_{1C}) and ^{13}C spin-spin relaxation time (T_{2C}), Horii et al[14] have separately recorded the spectra of the interfacial and amorphous components of

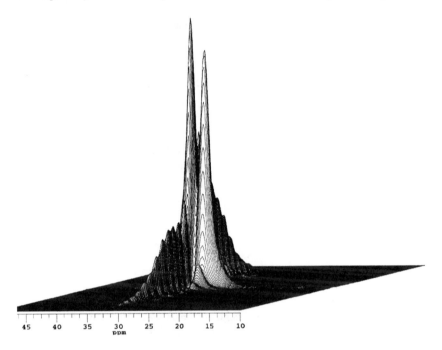

45	40	35	30	25	20	15	10
			ppm				

Figure 4. The solid state 2D Wideline Separation (WISE) spectra of the annealed EH copolymers, recorded at room temperature (CP time: 1 ms) without a mixing time before cross-polarization. The mobile component at 34 ppm is due to the interface.

the copolymers of ethylene and 1-hexene, prepared using metallocenes as catalyst, Then they resolved the fully relaxed DD/MAS spectrum into the three components, the crystalline, crystalline-amorphous interfacial, and rubbery amorphous components.

While ^{13}C-CP/MAS NMR spectroscopy gives valuable information about the molecular structure, molecular motion, morphology, and the existence of different phases of polyolefin samples, the amount of the phases as well as the dynamic behavior of the system can be probed by standard ^{1}H-NMR experiments. In addition, it is possible to correlate the dynamic information to specific sites within a system by the 2D-heteronuclear wide line separation (WISE) experiment.[11] The 2D WISE NMR experiment separates proton wideline spectra from different ^{13}C positions to establish a correlation of chemical structure and segmental mobility. This experiment is particularly useful for the investigation of miscibility and domain size of polymer blends and copolymer systems.[21] It is also a very useful method to detect matrix water in natural polymers, as shown recently for starch,[22] cellulose and cellulose/poly(vinyl alcohol) blends, [23] and wounded potato tissues.[24]

Figure 4 shows the solid state 2D WISE ^{13}C MAS NMR spectra of the annealed EH copolymer. The results of the ^1H line shape analysis indicated that while the ^1H line of amorphous region, the ^{13}C atoms of this region resonate at 31 ppm in the ^{13}C dimension, compose of one component Lorentzian lines, the ^1H line of order region, the ^{13}C atoms of the region resonated at 34 ppm, compose of two components. It should be noted that because of the spin diffusion process during the short cross polarization time, the proton of the amorphous region may partially diffuse to the ordered region via a spin diffusion process between the regions with different mobility. If the narrow line superimposed on the broad line of the ordered region comes mainly from the contribution of this proton spin diffusion process, the proton linewidth of the narrow component of the order region should be identical with the proton linewidth of the amorphous region.

The proton dimension lineshape analysis results indicate that the width of this narrow line at 34 ppm (^{13}C dimension) is 6.9 kHz, while the width of the pure amorphous signal at 30 PPM is 5.7 KHz. Therefore, the narrow line which is superimposed on the crystalline resonance at 34 ppm is assigned to the contribution from the interface between the crystal and amorphous regions. The significance of this result is that the solid state 2D NMR technique unambiguously shows the ordered all-trans conformational nature of the interface region based on the ^{13}C chemical shift detected from this region.

Branch Location

The amount of partitioning of branches in the crystalline and noncrystalline regions in the LLDPEs is also an important problem.[24-30] Based on the difference of proton relaxation times in the rotating frame, $^HT_{1\rho}$ and the spin diffusion phenomenon, morphology partitioning of branches in EP, EB, EH, EO copolymers, synthesized using Ziegler-Natta catalysts, has been reported.[15-18] It is reported that methyl branches are readily included in the crystalline region, while longer branches, such as butyl branches, are fully excluded from this region under equilibrium conditions. [24-29]

For the LLDPEs synthesized using metallocene catalysts, however, the $^HT_{1\rho}$ difference between the crystalline and amorphous phase is too small to detect the morphology partitioning of the branches. This is due to the fact that the LLDPEs prepared via metallocenes as catalysts have a relatively random distribution of branches and narrow molecular mass distribution. These features of these polyolefins are responsible for the unique properties, such as the excellent transparency and the unique morphology and molecular motion heterogeneity of the systems.

The crystalline and amorphous phase $^HT_{1\rho}$ values of EH copolymers in this study were determined to be 4.2 and 3.1 ms respectively. The difference of the relaxation times of the crystalline and amorphous regions is not adequate to selectively observe the spectroscopically pure crystalline and amorphous resonances, and it is also not adequate to study the branch partition. In order to study the branch location of the polyolefins synthesized using metallocene catalysts, we performed for the first time a dipolar dephasing experiment to directly follow the side chain decay behavior of the LLDPE samples with three different thermal histories. The dipolar dephasing technique allows the identification of resonances arising from different domains.[11] In this experiment, an extra delay, T_{DD}, is inserted at the end of the CP period and before acquisition in the presence of high power decoupling. A suitable delay will allow the magnetization of carbon nuclei from a rigid region, such as the crystalline phase, to decay to zero. The resulting spectrum is thus mainly due to carbon nuclei from mobile regions, such as the amorphous region above the glass transition temperature.

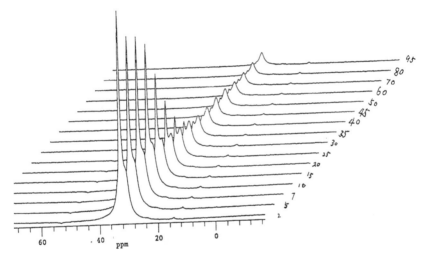

Figure 5. Stack plot of dipolar dephasing spectra of annealed EH copolymers with 16 different T_{DD}'s. The carbon resonance from the crystalline region decays fast. After a long waiting time, 90 μs, therefore, the crystalline signal is completely suppressed.

Figure 5 shows the ^{13}C dipolar dephasing CP/MAS NMR spectra of the annealed EH copolymers, EH-a, with 16 different T_{DD} values. The carbon resonance from the crystalline region decays fast. After a long waiting time, 60 μs, the crystalline signal is completely suppressed. After quantitative analysis of the decay curves of the crystalline and the amorphous resonances of the main-chain carbons, T_{DD} values of 15.5 and 50.3 s can be obtained for the crystalline region and amorphous region, respectively.

Due to the high resolution nature of the ^{13}C CP/MAS NMR spectra, the side-chain methyl group resonating at 14.1 ppm is well-resolved from the main-chain methylene carbon atoms. In order to investigate the branch partition for these LLDPE samples, the side-chain decay behavior of the copolymer of ethylene and 1-hexene was investigated for three different thermal histories. Due to the small amount and low intensity of the side chain, a large number of scans and long acquisition times were necessary to conduct the investigation. For each spectrum 1000 scans have been used; a total experimental time of 22 hours was required to acquire a set of dipolar dephasing spectra. After quantitative analysis of the decay curve of the side chain methyl resonance at 14.1 ppm, the T_{DD} values were obtained. Fig. 6 shows the plot of the peak intensity versus dipolar dephasing delay time, T_{DD}. The data indicate that while the T_{DD} data of main chain resonances in the crystalline region are about three times shorter than those in the noncrystalline region, the side chain T_{DD} shows a single exponential decay behavior. The T_{DD} values of the side chain for the SCBPE-q, SCBPE-a, SCBPE-sa, are 213, 123, 116 s, respectively. These results indicate that the morphologies of the side chains in the three samples are uniform, that is, the side chains are totally excluded from the crystalline regions.

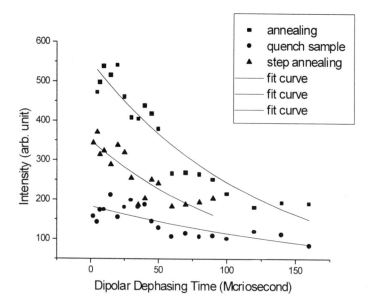

Figure 6. Plot of signal intensity of side chain methyl group versus delay time, TDD, in dipolar dephasing experiment of the copolymer of ethylene and 1-hexene with three thermal histories (For details, see experiment part). The solid lines indicate least-squares fits.

In-Situ Study of Molecular Crystallization Kinetics and Segregation

The overall crystallization kinetics and morphology of polyolefin copolymer s have recently reported in this laboratory.[10] The substantial difference in the behavior of the samples obtained from the homogeneous and heterogeneous melts indicated that a phase-separated melt exists which can be identified. DSC has been used to study molecular segregation during crystallization from the melt.[10] The crystallization kinetics data of the polyolefin samples obtained from different thermal histories can provide information about the heterogeneity, molecular and segmental segregation and phase morphology of the system. One key problem of this method is that it is very time consuming due to the small crystallinity change for each step. We recently developed a new solid state VT NMR method, which can detect the

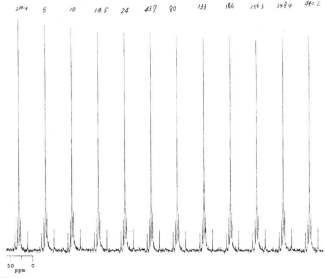

Figure 7. Stack plot of ^{13}C Bloch decay MAS spectra of EB sample at 70 °C with different annealing times after melting. The decay of the main chain amorphous signal can be quantitatively convert to the crystallinity change during this isothermal process.

molecular segregation kinetics in a much more efficient *in-situ* fashion. Figure 7 shows the isothermal ^{13}C Bloch decay stack spectra of the EB sample measured at 70 °C for different isothermal crystallization times quenched from the melt. The total experimental time is only about 10% as the time used in the DSC method[10]. The longer time required for the DSC method result from the necessary to prepare separate sample with different isothermal crystallization times from the melt, while in the NMR case the spectra of one sample with different isothermal crystallization times are acquired *in-situ* according to the sequence of the time elapsed. The method used to extract crystallinity data for this procedure is based on the fact that the carbon spin relaxation time of the amorphous region is very short, normally shorter than 1 second.[13] The quantitative amount of the amorphous phase based on after the first step of crystallization had been completed. This indicates that the crystals formed in the first step may serve as primary nuclei for the later crystallization steps. Furthermore, the crystals formed after the first step of crystallization may also change the dimensionality in their morphologies due to different SCB contents, and/or crystallization rates may deviate from linearity at constant temperatures, thereby substantially decreasing the Avrami exponents.

The other advantage of the variable temperature solid state NMR method is that the side chain information and the molecular motion information can be recorded at the same time under certain conditions. This information is valuable to study the evolution of the heterogeneous structure and to determine the long sequence distribution in the polyolefins copolymer systems

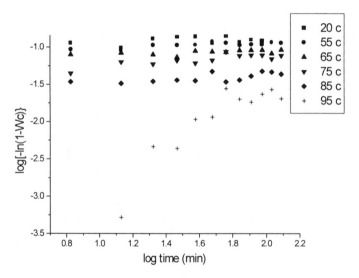

Fig. 8 Avrami treatments of a copolymer EO sample at six different temperatures via a multiple-step crystallization. The data were extracted from the variable temperature, solid state ^{13}C NMR experiments.

In conclusion, variable temperature solid state NMR is not only a useful and efficient analytical tool to determine the comonomer type and content in polyolefins as well as cross-linked polyolefins, but it is also a powerful and attractive approach to study the crystalline form, crystallinity, interfaces, and branch location of the samples as well. The advantages of the *in-situ* solid state VT NMR methods to rapidly extract the heterogeneous and molecular segregation kinetics information for polyolefins can be expected to continue opening new avenues for characterization and analysis of increasingly complex polyolefin copolymer systems.

REFERENCES

1. James, D. E. *Encycl. Polym. Sci. Eng.* **1986**, 6, 428
2. Balbontin, G.; Camurati, I.; Dallocco, T.; Finotti, A.; Franzese, R.; Vecellio, G. *Angew. Makromol. Chem.* **1994,** 219, 139
3. Liu, T. M.; Backer, W.E. *Polym. Eng. Sci.* **1992**, 32, 944
4. Channel, A. D.; Clutton, E. Q. *Polymer*, **1992**, 33, 4108
5. Hsieh, E.T.; Tso, C.C.; Byers, J.D.; Johnson, T.W.; Fu, Q.; Cheng, S.D. J. *Macromol. Sci.-Phys.* **1997**, 36, 615
6. Cheng, H. N. *Polym. Bull.* **1991**, 26, 325
7. Cheng, H. N. *Polym. Bull.* **1990**, 23, 589
8. Hatfield, G. R.; Killinger, W. E.; Zeigler, R. C. *Anal.Chem.* **1995**, 67, 3082.
9. Zeigler, R. C. *Macromol. Symposia* **1994,** 86, 213

10. Fu, Q. Chiu, F. Guo, M. McCreight, K. W. Cheng, S. Z. D. *J. Macromol. Phys.*, **1997**, 36, 1.
11. Schmidt-Rohr, K.; Spiess, H. W. *Macromolecules* **1992**, 25, 3273.
12. Yu, T.; Guo, M. *Progr. Polym. Sci.* **1990**, 15, 825.
13. Guo, M.; Fu, Q.; Cheng, S. Z. D. *Polym. Prepr. (Am. Chem. Soc. Div. Polym. Chem.)* **1997**, 38(2), 341.
14. Cheng, J.; Fone, M.; Reddy, V. N.; Schwartz, K. B.; Fisher, H. P.; Wunderlich, B. *J. Polym. Sci. Polym. Phys. Ed.* **1994**, 32, 2683.
15. Kuwabara, K.; Kaji, H.; Horii, F.; Bassett, D. C.; Olley, R. H. *Macromolecules* **1997**, 30, 7516
16. Inoue, D.; Kurosu, H.; Chen, Q.; Ando, I. *Acta Polym.* **1995**, 46, 420.
17. Morin, F.G.; Delmas, G.; Gilson, D.F.R. *Macromolecules* **1995**, 28, 3248.
18. Kitamaru, R.; Horii, F.; Zhu, Q.; Bassett, D.C.; Olley, R.H. *Polymer* **1994**, 35, 1171.
19. Guo, M. Manuscript in preparation.
20. Kitaamaru, R.; Horii, F.; Murayama, K. *Macromolecules* **1986**, 19, 636.
21. Guo, M. *Trends Polym. Sci.* **1996**, 4(7), 238
22. Kulik, A. S.; de Costa, J. R. C.; Haverkamp, J. *J. Agr. Food Chem.* **1994**, 42, 2803.
23. Radloff, D.; Boeffel, C.; Spiess, H.W. *Macromolecules* **1996**, 29, 1528
24. Yan, B.; Stark, R.E. *Macromolecules* **1998**, 31, 2600
25. Perez, E.; Vanderhart, D. L. *J. Polym. Sci. Polym. Phys. Ed.* **1988**, 26, 1979.

26.   Perez, E.; Vanderhart, D. L.; Crist, B.; Howard, P. R. *Macromolecules* **1987**, *20*, 78.
27.   Perez, E.; Vanderhart, D. L. *J. Polym. Sci. Polym. Phys. Ed.* **1987**, *25*, 1637.
28.   Laupretre, F.; Monnerie, L.; Barthelemy, L.; Vairon, J. P.; Sauzeau, A.; Roussel, D. *Polym. Bull.* **1986**, *15*, 159.
29.   VanderHart, D. L.; Pérez, E.; *Macromolecules* **1986**, 19, 1902.
30.   Bassett, D. C. *Developments in Crystalline Polymers* 2; Applied Science: London, **1988**; pp 67-103.

Chapter 12

Determination of Trimethylaluminum and Characterization of Methylaluminoxanes Using Proton NMR

Donald W. Imhoff[1], Larry S. Simeral[2,3], Don R. Blevins[2],
and William R. Beard[1]

[1]Process Development Center, Albemarle Corporation, P.O. Box 341,
Baton Rouge, LA 70821
[2]Albemarle Technical Center, P.O. Box 14799, Baton Rouge, LA 70898

New methods for characterization of methylaluminoxanes (MAO) and determination of trimethylaluminum (TMA) using proton NMR are reported. Addition of excess perdeuterotetrahydrofuran to sharpen and shift the TMA followed by curve fitting to remove residual overlap between the TMA and MAO peaks allow accurate, precise and rapid quantitation of TMA and MAO in solution. The analysis also provides determination of all the solution components of MAO preparations and the number of methyl groups per aluminum in MAO when combined with an independent determination of the aluminum content. Accurate number average molecular weights for MAOs are determined from freezing point depression data corrected for TMA and other small molecule concentrations obtained using the NMR method.

Methylaluminoxane (MAO) is an important co-catalyst for the oligomerization and polymerization of a variety of monomers. *(1-7)* Despite the commercial importance of MAO and the considerable effort directed at its characterization, the structure of MAO remains unclear. *(7-10)* The presence of multiple equilibria and alkyl group exchange prevent easy direct structural characterization. Methylaluminoxane is usually written as the oligomer formula $[-Al(CH_3)O-]_n$ and the literature contains many postulated structures, including chains, cages, and rings of various sorts. *(7-10)* Further, residual trimethylaluminum (TMA) in MAO preparations clouds structural interpretations. *(7-11)*

Controlled hydrolysis of TMA in toluene or other hydrocarbon solvent to form MAO leaves residual TMA in the product solution. *(7,10,11)* The TMA in MAO is present as free and bound species according to the equilibrium:

$$MAO + Al(CH_3)_3 \leftrightarrows (MAO) \cdot Al(CH_3)_3 \tag{1}$$

Commercial MAOs are known to contain significant quantities of TMA which can have important effects on the catalytic activity of the MAO. *(10-11)* The amount of residual TMA is an important parameter for characterizing MAO and for controlling MAO product quality.

Several approaches to determine residual TMA have been explored in the

[3]Corresponding author.

literature: a) determination of the volatile aluminum content from MAO solutions; b) pyridine and t-butanol titration of total TMA *(12)*; c) P-31 NMR spectroscopy *(11)*; d) C-13 NMR spectroscopy *(13)*; and e) proton NMR. *(14)* The determination of volatile aluminum content for TMA is inaccurate since only the "free" TMA is volatilized. Pyridine titration assumes that pyridine complexes only with TMA and not with MAO. However, side reactions of pyridine with isolobal gallium sulfide and with other aluminoxanes have been reported. *(15-16)* Pyridine appears to complex with sites on the MAO yielding a high TMA analysis. *(11, 13)* Heteronuclear NMR methods (P-31 and C-13) use spectroscopic probe molecules to determine the residual TMA. These heteronuclear NMR methods indirectly determine the TMA using the chemical shift of the appropriate nucleus in the probe molecule. Since the probe molecules are typically Lewis bases, interaction of the probe or reporter molecule with MAO cannot be unequivocally ruled out in many cases. Further, heteronuclear chemical shifts are often very sensitive to solvent, concentration and temperature effects. None of the above approaches address the characterization or analysis of MAO or any other species in solution.

Proton NMR has several advantages in analysis of MAO products. The proton NMR spectrum shows NMR resonances for all proton containing species present in the sample, including TMA, MAO, solvent, additives, and contaminants even at very low concentrations (<0.1 wt%). A proton NMR analysis is a direct analysis of the components in solution. Further, the high sensitivity of proton NMR makes the analysis very rapid and, therefore, useful for quality control analysis. Previous reports ruled out the accurate and precise determination of TMA content using proton NMR because of the fluxional nature of MAO and severe overlap of the TMA and MAO Al-methyl group resonances. *(10,11,14)* We report here a new approach to the proton NMR analysis of TMA in MAO solutions which effectively eliminates the TMA/MAO spectral overlap and significantly narrows the methyl group resonance of TMA, allowing good quantitative determination of TMA and the other species in solution. The method we demonstrate is both rapid and precise.

Once the amount of TMA is accurately determined, some insight into the constitution of MAO can be determined directly from the proton NMR spectra. We show here that the proton NMR data combined with an independent determination of the aluminum content of the sample provides characterization of the MAO through determination of the number of methyl groups per aluminum atom in MAO. This allows further refinement of the molecular formula of MAO and helps set limits for possible structural elements. Determination of the methyl groups per aluminum begins to provide a basis for analytical comparisons of different MAOs. Further, the development of a rapid and accurate method for both MAO and TMA analysis eliminates the need for multiple labor and time intensive classical chemical methods widely used for MAO analysis.

Experimental

All sample preparations were performed under dry nitrogen in an inert atmosphere glove box. Perdeuterobenzene, perdeuterotetrahydrofuran (THF-d_8), perdeuteroacetonitrile and perdeuterodioxane were obtained from Cambridge Isotopes at 99.6 % or higher deuterium. The solvents were dried using sodium/potassium alloy and/or activated basic alumina. TMA was obtained from Albemarle Corporation and MAOs were obtained from Aldrich and Albemarle. Most of the spectra were obtained from samples of MAO in toluene to which the deuterated solvent was added for field/frequency lock. The toluene resonance was assigned to 2.09 ppm. Five millimeter NMR tubes were dried at 110 °C for at least one hour before transfer to the glove box for sample preparation.

Proton NMR spectra were obtained at 400 MHz on a Bruker DPX400 or a Bruker/GE Omega 400WB instrument and at 300 MHz on a Bruker/GE QE-300 instrument. Proton NMR spectra were obtained using 5 mm quad nucleus probes on the DPX400 and QE-300 instruments and a 5 mm inverse probe on the Omega 400WB instrument. Typical data collection parameters provided a digital resolution for the spectrum (real data points) of 0.2-0.3 Hz per data point for accurate quantitation of integrals. Relaxation times were measured using the inversion recovery method.

Considerable attention was paid to the details of the NMR instrument operation and parameters to ensure good quantitative results. We used 30 deg pulse widths and pulse intervals of 3-5 times the longest T_1 (recovery of >99% of signal intensity for a 30 deg pulse) of the components to be measured to ensure complete relaxation. The real data digital resolution used was typically 0.2-0.3Hz per data point which provided adequate digitization for all the resonances of interest. The spectral response of the instrument audio filtering was linear over the observed spectral region. This was assured by proper choice of sweep width and filter droop correction or by full use of linear digital filters (DPX400). The number of acquisitions was typically 4-8 at 400 MHz and 32-64 at 300 MHz, or enough to make the ^{13}C satellites of the toluene at least 10 x the height of the noise. Prior to integration careful peak phasing and correction for any baseline dc-offsets or anomalies were performed. Integrals were taken over the peak regions broad enough to include the ^{13}C satellites. Curve fitting to remove the broad MAO resonance for integration of the TMA peak was performed using multi-factor polynomial curve fitting routines provided with the NMR instrument software.

Sample preparation differed for the two approaches discussed in this paper. For the internal standard approach about 0.3 g of the MAO sample solution in toluene along with 0.1 g mesitylene (internal standard) were accurately weighed into a vial. After thoroughly mixing, one part by volume of this mixture was mixed with four parts THF-d_8 directly in the NMR tube and again thoroughly mixed. For the normalization approach no weighings were required. One part of the MAO solution was added directly into a 5 mm NMR tube containing four parts of THF-d_8. The NMR measurements were made after thorough mixing of the sample. For both approaches substitution of perdeuterodioxane or perdeuteroacetonitrile for the THF-d_8 yielded the same results.

Pyridine titration was performed by titrating a standardized stock solution of pyridine in toluene containing phenazine dye with the MAO solution.

Aluminum analyses were performed with equivalent results using two methods. Wet chemical total aluminum determination was performed by dissolving an aliquot of MAO solution in bis-2-methoxyethyl ether, followed by alcoholysis. After acidification and addition of dithizone indicator, an EDTA/zinc acetate back titration was performed. Inductively coupled plasma atomic emission (ICP-AE) spectroscopy was used to determine the aluminum in aqeous solutions obtained after hydrolysis followed by dissolution with hydrochloric acid. The organics were volatilized by boiling prior to introduction of the sample into the ICP.

Neutron activation analysis for total aluminum and oxygen in MAO was performed at the Texas A & M University Center for Characterization and Analysis, College Station, TX.

The number average molecular weight (Mn) of MAO was determined using a freezing point depression cryoscopic technique with dry 1,4 dioxane as the solvent. Benzil was used as the standard for determination of the cyroscopic constant. All operations were performed in a drybox under nitrogen. An accurate weight of MAO (about 2 grams) was dissolved in 50 mL of dioxane and 40 mL of the solution was poured into a glass, jacketed, cylindrical container. A few milligrams of crushed glass was added to provide a nucleating surface to aid in the initiation of crystallization, thus preventing excessive super-cooling. The solution was stirred continuously with a magnetic stirrer, and its temperature was monitored using a platinum resistance thermometer coupled to a PC with temperature data are recorded every 5 seconds. A liquid cooling medium at ^{0}C was pumped through the container's jacket. The freezing point was determined as the highest temperature reached after super-cooling.

The solid MAO sample for Mn analysis was prepared by high-vacuum distillation of volatile components (toluene and TMA) into a -196 °C trap. Proton NMR analysis (as described in this section and in the Results and Discussion) was performed to accurately determine the residual levels of toluene, TMA and process oil. Typically, these dried solids contained about 1wt% toluene, 1.5 wt% process oil and 10 wt% TMA.

NMR analysis of the MAO solid dissolved in the dioxane showed no significant difference from the NMR analysis of the original MAO solution in toluene other than the expected loss of toluene and TMA in the drying process. Significant reaction of dioxane with MAO other than complexation of available Lewis acid sites is not expected. The freezing point depression results were corrected for the small molecular weight components todetermine the "real" Mn for MAO.

Results and Discussion

Aluminum alkyls complex reversibly with many Lewis bases to form adducts. (11,13,17) Lewis bases, such as pyridine and tetrahydrofuran (THF), are known to cleave alkyl bridged dimers to form the adduct:

$$(AlR_3)_2 + 2B \leftrightarrows 2 (B \cdot AlR_3). \tag{2}$$

Addition of excess THF to TMA solutions shifts the equilibrium to the right. In the proton NMR spectrum of TMA in aromatic hydrocarbon solvents, addition of excess THF narrows the TMA resonance significantly and shifts the TMA resonance upfield by about 0.5 ppm. Exchange of TMA between the free and complexed states is rapid and a single sharp peak is observed. We use this information to aid the analysis of TMA in MAO solutions. Figure 1a shows the proton NMR spectrum of a 30% MAO in toluene solution diluted in perdeuterobenzene to about 5% MAO. The MAO resonance is very broad and featureless. The broad and overlapping TMA peak (6-10 Hz linewidth at half-height) is superimposed on that of the MAO. Figure 1b shows the spectrum of the same

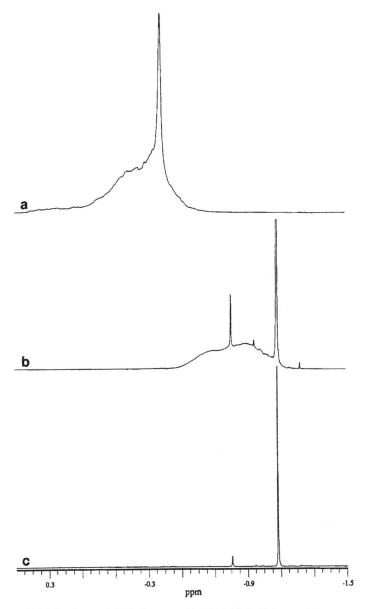

Figure 1. 400 MHz Proton NMR Spectra of 30% MAO in Toluene, 0.5 to -1.5 ppm:
a) Diluted in perdeuterobenzene only. The TMA peak is at -0.37 ppm, while the
broader feature, 0.5 to -0.7 ppm, is the MAO.; b) Diluted in 4 volumes of THF-d$_8$.
The TMA peak is at -1.08 ppm. This spectrum is expanded vertically to see the broad
MAO peak, -0.3 to -1.3 ppm.; and c) Spectrum b) with the MAO removed via curve
fitting. The small peak at -0.8 ppm is a low molecular weight species or end-group
from the MAO. The T$_1$ of this peak closely matches the MAO and is much shorter
than that of TMA.

sample to which excess THF-d$_8$ (4:1 volume to volume, or about 4:1 molar ratio of THF to total aluminum) has been added. The TMA resonance is now sharp (0.8-1 Hz linewidth at half-height) and nearly resolved from the MAO. Further, the MAO resonance also shifts slightly upfield. The optimum shift and narrowing for TMA is observed at about a 4:1 v/v ratio of THF to MAO solution. Figure 1c shows the results of curve fitting a "baseline correction" to the MAO resonance on both sides of the TMA peak, leaving the TMA peak completely resolved and easily integrated. The addition of excess THF-d$_8$ followed by curve fitting to remove any residual overlap between the TMA and MAO peaks provides the basis for separation of the TMA and MAO resonances for quantitation. Use of acetonitrile or dioxane provides equivalent results. Excess THF-d$_8$ was used for the results reported here.

There are two approaches for proton NMR which can be used for the quantitation of TMA in MAO after the problem of peak overlap is solved using the THF-d$_8$ and curve fitting procedure. One approach is to use a weighed internal standard and calculate the weight percent of TMA from the weight of sample and standard plus the integrals of standard and TMA. The second approach assumes all the solution species are known and the results are normalized to 100%. The internal standard approach is most useful for modified and derivatized MAOs and where insolubles are present. We report and discuss results of both methods.

Some care must be taken in choosing an internal standard for highly reactive chemicals, such as MAO and TMA. The internal standard must be inert toward TMA and MAO and have NMR peaks which do not overlap with any of the species in solution, including the solvent toluene. Mesitylene, diphenylmethane and 1,2-diphenylethane are good choices. Chlorinated hydrocarbons must be avoided because of their reactivity with TMA and MAO.

Table I shows the T$_1$ values for 30% MAO in toluene with added mesitylene diluted 1:4 in THF-d$_8$. Clearly, the aromatic protons of toluene have the longest relaxation time. The relaxation times may seem unusually long for proton NMR. However, note that the MAO samples were prepared in the absence of oxygen and therefore, a significant source of relaxation has been removed. Despite the length of the T$_1$s, the sensitivity of the proton NMR experiment is such that data are obtained in less than 15 minutes for TMA analysis.

Table I. Proton NMR Relaxation Times of MAO Solution Species.

Species	T$_1$ (sec)
TMA	7.6
MAO	0.7
CH$_3$ (toluene)	10.0
Aromatic (toluene)	26
Hydrocarbon	4.0
CH$_3$ (mesitylene)	5.7
Aromatic (mesitylene)	18.4

Table II compares the result of TMA determination using proton NMR and mesitylene as an internal standard with pyridine titration for the same samples. Note that the TMA concentration determined using proton NMR is substantially less than the pyridine titration results. High TMA values from pyridine titration are expected. *(11,13)* To determine whether all of the TMA is detected by this proton NMR procedure, we added additional known amounts of TMA to the samples in a "standard additions" approach. The added TMA plus TMA already present was completely recovered by the method. This work and others *(11,13,15,16)* point to high pyridine titration numbers from interaction of the pyridine with Lewis acid sites on the MAO. The pyridine appears to interact with sites on the MAO with reactivity similar to that of TMA.

Table II. Wt% of TMA in MAO Solutions Using Proton NMR with an Internal Standard[1]

Sample	TMA from NMR[2]	TMA from pyridine titration
1	5.2	7.3
2	2.1	3.3
3	1.7	2.3

1) Dry mesitylene was used as the internal standard here. 2) The estimated precision is 4% relative at one standard deviation.

It is important to recognize that more information can be obtained from the proton NMR analysis of MAO solutions. Just as we have measured the amount of TMA, the observed solvent (toluene), and residual amounts of process oil (a minor component left from the manufacturing process) can be quantitated. The only component remaining is the MAO which can readily be calculated by difference. Direct measurement of the methyl groups of the MAO moiety is also possible. Subtraction of the TMA peak area from the entire integral for all methyl groups on aluminum, that is, the region between 0.0 and -1.3 ppm, yields the area of the MAO methyls. Quantitation of the MAO based on the methyl groups measured requires insight into the MAO unit formula. When we assumed a MAO repeating unit formula of $[-Al(CH_3)O-]_n$ with a ratio of 1:1:1 for CH_3's, Al, and O's, as is often proposed, *(7)* we found inconsistences between the MAO as measured using the NMR difference approach and the aluminum content as determined by wet chemical methods. The unit formula that fit all the data had a ratio of about 1.4:1:0.8 for methyls, aluminum, and oxygen, respectively.

From these results arose a simplified normalized NMR analytical method. This approach requires no internal standard, no weighings (for the NMR portion), and can be completed in less than 15 minutes. The amounts of TMA, MAO, toluene, and process oil are determined along with a measure of the number of methyl groups per aluminum atom in the MAO structure. Additionally, the number of millimoles of gas (methane) per gram of sample and moles of gas per mole of aluminum can be calculated. An independent method for the determination of the aluminum content is required. Accurate aluminum measurements are routinely made on MAO solutions by hydrolysis followed by wet chemical or ICP-AE spectroscopy methods. This aluminum content combined with the proton NMR measurements provide the complete analysis. This NMR procedure eliminates the need for the conventional tedious and very time consuming gas evolution methods typically employed in the MAO industry.

Three assumptions are necessary in this NMR normalization approach:

1) All the aluminum determined independently (ICP-AE or wet chemical) is present and is observed by the methyl groups of MAO or TMA. Species like Al_2O_3 , $AlCl_3$, $Al(OH)_3$, and free Al would not be seen by ^{1}H-NMR; these are not expected and most would be insoluble in toluene solution. Further, we find no evidence for such compounds in typical MAO preparations.

2) All the components present are observed by ^{1}H-NMR. Normalization methods require all components of significance to be measured. No non-proton materials, for example totally halogenated solvents and/or reactants, are used in the manufacturing process and indeed none are expected. Very minor components which may not be detected would be insignificant in the normalization calculations.

3) The valences of aluminum, oxygen and methyls in MAO are satisfied by the formula:

$$[-(CH_3)_X \, Al \, O_{((3-X)/2)} -]. \qquad (3)$$

Another plausible structural component of MAO is the hydroxyl group. Such groups have not been reported previously in commercial MAOs and we find that treatment of MAO with various hydride reducing agents, such as lithium triethyl borohydride, yields insignificant amounts of hydrogen, indicating that very few, if any, hydroxyl groups were present. Additionally, neutron activation analyses for several MAO samples gave Al/O ratios about 1:0.8 in MAO, consistent with the results obtained by this NMR procedure. Further, we find no evidence for methoxy groups in the proton NMR spectra of MAO samples unless they have been exposed to oxygen. Thus, we conclude that MAO is comprised of Al-methyl and Al-oxygen groups only.

Weight percentages are calculated according to the usual normalization methods. However, before these calculations are done, the number of methyl groups (X in assumption 3 structure) per aluminum of the MAO moiety must be determined. The aluminum content from independent analysis, plus the measured integrals for each component of concern, provide the necessary equations to solve for a discreet value for X (see Appendix 1). The molecular weight of the average MAO unit then is:

$$MW_{MAO} = 51.0 \text{ g/mol} + 7.0 * X \text{ g/mol}. \qquad (4)$$

Table III shows the results obtained for three typical MAO samples, labeled ALB30, ALB20, and ALB10, produced by Albemarle Corporation. The Aldrich 10% sample was obtained from Aldrich Chemical, but the producer of this material is not known. Sample ALB30 was diluted with dry toluene to produce sample ALB30D. The calculated results based on the dilution factor are shown in brackets below the results obtained. The good agreement clearly shows the method is indeed independent of the concentration of MAO and TMA. A precision study was performed using sample ALB30. This sample was prepared thirteen times and the NMR data were obtained at 400 MHz on the Bruker DPX400 by three operators on four different days. The percent relative standard deviation data are shown in parenthesis below the sample values.

The precision and consistency of the data using the normalized proton NMR

Table III. Data from the Normalized Wt% Analysis of MAO Solutions Using Proton NMR

Sample	TMA Wt%	TMA As % Al	MAO	Oil	Toluene	CH$_3$ per Al on MAO	mmol gas per gm of sample[1]	moles gas per mole of Al[1]
ALB30	5.17 (1.1)[2]	14.1 (1.1)	26.51 (0.21)	0.50 (2.2)	67.81 (0.03)	1.41 (0.31)	8.29 (0.27)	1.63 (0.25)
ALB20	3.50	14.5	17.48	0.34	78.69	1.41	5.51	1.64
ALB10	1.78	14.0	9.19	0.21	88.73	1.44	2.93	1.66
Aldrich 10%	4.01	30.4	7.79	0.85	87.35	1.48	3.55	1.94
ALB30D[3]	1.24 [1.21][4]	14.5 14.3	6.17 [6.20]	0.13 [0.12]	92.46 [92.48]	1.40 [1.40]	1.93 [1.94]	1.63 [1.63]

1) Values calculated from NMR data and the aluminum analysis.
2) The values shown in ( ) are the precision of the analysis in % relative standard deviation. The precision was determined from thirteen replicates obtained by three operators over four days.
3) The sample ALB30D is sample ALB30 diluted with dry toluene a factor of 4.277 times.
4) The values in [ ] are the calculated numbers based on the dilution factors.

analysis are excellent. The relative standard deviation is about one percent at 5% TMA. The good agreement between calculated and found values for the diluted sample shows the method is indeed independent of MAO and TMA concentrations. The precision of the method is very good over a wide range of TMA concentrations and is good enough to allow detection of relatively subtle changes in MAO solution composition.

All the results on MAO samples, including that from Aldrich, show the number of methyl groups on the aluminum of MAO (excluding TMA) is in the range of 1.4 to 1.5, while oxygen ranges from 0.75 to 0.80 atoms per aluminum. This method provides a characterization of the MAO moiety in toluene solution showing, on the average, 40 to 50 percent of the aluminum atoms have two methyl groups attached. This finding is contrary to many of the models found in the literature which predict or show a 1:1:1 ratio for CH_3, Al, O. These "excess" methyl groups may contribute to MAO's efficacy as a cocatalyst. It is suggestive that they may provide a site or sites of activated methyl groups available for facile exchange with halogens or methyl groups on a metallocene catalyst. An MAO with 1:1:1 ratio may be a very poor cocatalyst. These results suggest the formula for MAO is best represented as

$$[-(CH_3)_{1.4 \text{ to } 1.5} \text{ Al } O_{0.80 \text{ to } 0.75} -]_n. \tag{4}$$

It is interesting to note that the Aldrich 10% MAO in the table above contains considerably more TMA than the comparable Albemarle MAO sample. The ALB samples contained 14-15% of the total aluminum as TMA, while the Aldrich contained about 30% as TMA. The role of TMA in MAO as a cocatalyst is not completely understood, but TMA activity as a cocatalyst is very much lower than MAO activity and ethylene polymerization productivity can be lower with increasing TMA concentrations. *(18)* Whatever TMA's role, this method serves to accurately measure its presence and can be used to control its level in commercial MAO.

Another important application of accurate data for the composition of MAO preparations is determination of MAO number average molecular weight (Mn) from freezing point depression data. The accurate determination of colligative properties, such as freezing point depression, requires a knowledge of the concentration of all the solution components. This is especially critical for low molecular weight components, such as TMA, process oil and toluene. Table IV shows some example apparent and corrected Mn for MAO in dioxane. Clearly, a knowledge of the complete composition of MAO materials is necessary for accurate analysis of Mn. Even "dried" solids of MAO contain measureable amounts of toluene, TMA and process oil.

Figures 2, 3 and 4 show possible model structures for MAO species based on the NMR and Mn data combined with available literature. These structures are based on well characterized model compounds, have the empirical formula described in this report, and have molecular weights in the 700-1200 amu range. It must be emphasized that many other structures could be drawn that also fit the available data. The above formula suggested for MAO from the NMR data combined with Mn data provides boundary conditions for developing structural models.

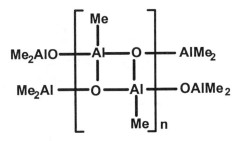

Figure 2. A structure that almost fits the data. For n=3, the formula is $Me_{14}Al_{10}O_8$ having Mn=608 amu, Me/Al=1.40, and O/Al= 0.8. For n=4, the formula is $Me_{16}Al_{12}O_{10}$ having Mn=724 amu, Me/Al=1.33, and O/Al=0.83.

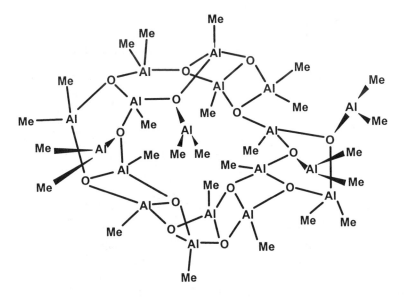

Figure 3. A modified "Sinn" oligomer.(*19*) The formula is $Me_{26}Al_{18}O_{14}$ having Mn=1100 amu, Me/Al=1.44, and O/Al=0.78.

Table IV Apparent and Corrected Molecular Weights for Typical MAO Samples.[1]

Sample No.	Apparent Mn (amu)	Corrected Mn (amu)
1	439	701
2	402	817
3	386	931
4	432	1110
5	367	786

1) Mn determined from freezing point depression in dioxane solution as described in the Experimental Section. Precision is about +/- 5% relative. The corrected Mn is calculated with corrections for TMA, toluene, process oil and MAO concentrations determined for the solid sample in perdeutero THF using NMR and the aluminum weight percent from ICP-AE.

Conclusions

We have described here proton NMR methods to accurately measure the TMA content in MAO preparations. The normalized NMR method combines the aluminum content from an independent method to shed new insight into the MAO formula itself. The measurement of the number of methyl groups on aluminum in MAO is easily obtained. The NMR data allow the calculation of the amount of gas (methane) per gram of sample or per mole of aluminum that would be evolved if a lengthy and tedious gas evolution method were done (see Table III).

This method provides a powerful means to monitor the MAO manufacturing process and the final MAO product. Although it by no means gives a complete description of the structure of MAO, it gives a new insight into its formula and structural elements. The method provides a time efficient and cost effective means to follow the level of TMA and help assure a consistent MAO product. Further, the methods help provide accurate and reliable number average molecular weights for MAO.

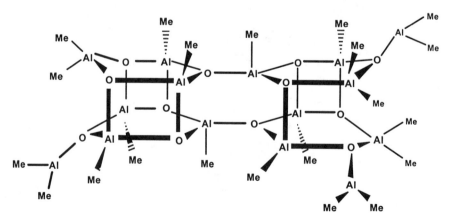

Figure 4. Fused "Barron" cages(*20*) reacted with TMA. The formula is $Me_{21}Al_{15}O_{12}$ having Mn=912 amu, Me/Al=1.40, and O/Al=0.80.

Acknowledgments

The authors gratefully acknowledge the help of David Bugg (ICP-AE methods and analyses), Sherri Joubert (helpful discussions plus aluminum, gas evolution and pyridine titration determinations), James Peel (sample analysis and dry solvents), Sam Sangokoya (helpful discussions), and Karl Wiegand (helpful discussions).

Literture Cited

1. H. Sinn, W. Kaminsky, *Adv. Organomet. Chem.* **1990**, *18*, 429.
2. H. Sinn, W. Kaminsky, H. J. Vollmer, R. Woldt, *Angew. Chem. Int. Ed. Engl.* **1980**, *93*, 390.
3. J. Bliemeister, W. Hagendorf, A. Harder, B. Heitmann, T. Schimmel, N. von Thienen, K. Urlass, H. Winter, O. Zarnicke, in *"Ziegler Catalysts"*, G. Fink, R. Mulhaupt, H. H. Brintzinger eds., Springer-Verlag, New York **1995**, pp.57-82.
4. E. J. Vanderberg, *ACS Symp. Ser.* *1992, 496*, 222.
5. V. R. Gupta, M. Ravindranathan, *Macromol. Chem. Phys.* **1996**, 197, 1937 ().
6. D. Wilson, *Polymn. Int.* **1996**, *39*, 235.
7. S. Pasynkiewicz, Polyhedron **1990**, *9*, 429.
8. C. H. Harlan, S. G. Bott, A. R. Barron, *J. Amer. Chem. Soc.* **1995**, *117*, 6465.
9. C. Janiak, R. Bernhard, R. Voelkel, H.-G. Braun, *J. Polym. Sci. Part A* **1993**, *31*, 2959.
10. I. Tritto, M. C. Sacchi, P. Locatelli, S. X. Li, *Macromol. Chem. Phys.* **1996**, *197*, 1537.
11. A. R. Barron, *Organometallics* **1995**, *14*, 3581.
12. D. E. Jordan, *Anal. Chem.* **1968**, *40*, 2150.
13. G. M. Smith, D. B. Malpass, S. N. Bernstein, D. Jones, D. J. Monfiston, J. Rogers, B.-M. Su, S. W. Plmaka, C. F. Tirendi, C. C. Crapo, D. L. Deavenport, J. T. Hodges, R. H. Ngo, C. W. Post, C. S., Robbitt, in *SPE Polyolefins X International Conference*, SPE Houston, **1997**, pp 49-69.
14. L. Resconi, S. Bossi, L. Abis, *Macromolec.* **1990**, *23*, 4489.
15. M. B. Power, J. W. Ziller, A. R. Barron, *Organometallics* **1993**, *11*, 2783.
16. M. R. Smith, J. M. Mason, S. G. Bott, A. R. Barron, *J. Amer. Chem. Soc.* **1993**, *115*, 4971.
17. A. R. Barron, *J. Chem. Soc., Dalton Trans.* **1988**, 3047.
18. E. Gianetti, G. M. Nicoletti, R. Mazzocchi, *J. Polymer Sci. Part A* **1985**, *23*, 2117.
19. H. Sinn, *Macromol. Symp.* **1995**, *97*, 27.
20. M. Watanabe, C. J. Harlan, A. R. Barron, *"Olefin Polymerization Using Well-Defined Co-Catalysts."* 214th American Chemical Society National Meeting, Las Vegas, NV, September 7-11, 1997, Division of Inorganic Chemistry, Paper #441.

Appendix 1. Normalization Equations.

Let A = area for toluene $[\dfrac{\text{Aromatic H's}}{5} + \dfrac{CH_3\text{'s}}{3}] / 2$

B = area for TMA/9
C = area for MAO/3 (after subtracting TMA area)
D = area for mineral oil/2
K = wt % Al/100*27.0
X = number CH_3's on Al in MAO
MW = molecular weight of the average MAO unit

Molecular weights: Toluene = 92.1; TMA = 72.1; Mineral Oil = 14.0

(1) wt % Toluene $= \dfrac{A*92.1}{[A*92.1 + B*72.1 + C*MW/X + D*14.0] = Z} *100$

(2) wt % TMA $= \dfrac{B*72.1}{Z} * 100$

(3) wt % Mineral Oil $= \dfrac{D*14.0}{Z} * 100$

(4) wt % MAO $= \dfrac{C*MW/X}{Z} * 100$

The weight percent Al is distributed between TMA and MAO as follows:

$$\dfrac{\text{wt \% Al}}{100} = \dfrac{72.1\,B}{Z} * \dfrac{27.0}{72.1} + \dfrac{MW*C}{ZX} * \dfrac{27.0}{MW}$$

rearranging gives:

(5) $K = \dfrac{\text{wt \% Al}}{27.0*100} = \dfrac{B}{Z} + \dfrac{C}{ZX}$

and now solving for X gives:

(6) $X = \dfrac{C - KC*MW}{KA\,92.1 + KB\,72.1 + KD\,14.0 - B}$

Consider MAO unit $(CH_3)_x\text{-Al-O}_{(3-x)/2}$

$$MAO(MW) = (mwAl) + X(mwCH_3) + (\dfrac{3-X}{2})(mwO)$$

$$MW = 27.0 + X(15.0) + (\dfrac{3-X}{2})16.0$$

(7) $MW = 51.0 + 7.0\,X$

Substituting this value for molecular weight into equation (6) above and solving for X gives:

$$(8) \quad X = \frac{C - KC*51.0}{KA*92.1 + KB*72.1 + KD*14.0 - B + 7.0*KC}$$

Now that the number of CH_3's on Al in MAO and the unit molecular weight is established (equation 7), the weight percentages of TMA, MAO, toluene, and mineral oil can be calculated with equations 1-4 above.

Furthermore, the mmoles of gas per gram of sample and the moles of gas per mole of Al are readily calculated

$$(9) \quad \text{mmoles gas/gr. sample} = [\frac{(wt\% MAO)*X}{100*MW} + \frac{(wt \% TMA)*3}{100*72.1}]*1000$$

$$(10) \quad \text{mmoles Al/gr. sample} = \frac{wt \% Al*1000}{100*27.0}$$

$$(11) \quad \text{moles gas/moles Al} = \text{equation (9)/equation (10)}$$

INDEXES

Author Index

Subject Index

A

Anionic functionalization, incorporating *p*-methylstyrene into polyolefins, 120–121

Anionic graft-from reactions, lithiated polyethylene powder with styrene or *p*-methylstyrene, 122–123

Asymmetric chains
enantiomorphic-site control polymers, 131
See also Poly(propylene)s, highly isotactic

Asymmetric/symmetric chain distribution. *See* Poly(propylene)s, highly isotactic

B

Benz indenyl ligand. *See* Zirconocene complex with benz indenyl ligand

Bis(indenyl) metallocene complexes. *See* Siloxy substituted group IV metallocene catalysts

Bis(tetrahydroindenyl) metallocene complexes. *See* Siloxy substituted group IV metallocene catalysts

Branch location
partitioning in crystalline and noncrystalline regions of LLDPEs, 170
signal intensity of side chain methyl group versus delay time, 172*f*
stack plot of dipolar dephasing spectra of annealed ethylene/1-hexene (EH) copolymer, 171*f*
See also Linear low-density polyethylene (LLDPE)

Bridged complexes. *See* Zirconocene complex with benz indenyl ligand

Butadiene, polymerization with neodymium catalysts, 19–20

Butadienes, substituted
2,4-hexadiene polymers, 27
polymerization with neodymium systems, 21, 23–27
polymer microstructure, 21, 23–25
polymer molecular weight, 25
polymers from terminally substituted monomers, 25, 27
X-ray powder spectra of polymers from terminally substituted monomers, 28*f*

C

Catalyst activity, conventional neodymium systems, 16

Catalyst preparation, conventional neodymium systems, 16

Cerium catalysts, polymerization of 1,3-dienes, 15

Chain distribution. *See* Poly(propylene)s, highly isotactic

Chain sequence distributions, Markovian statistics, 131–132

Cis content of polymer, conventional neodymium systems, 16–17

Comonomer types
determination and contents in melt state, 166–167
fast determination of ethylene copolymers at room temperature, 165, 166*f*

Copolymerization. *See* Monocyclopentadienyltitanium/MAO catalysts; Zirconocene complex with benz indenyl ligand

Copolymers
poly(ethylene-*co*-*p*-methylstyrene), 106–108
See also Ethylene copolymers, amorphous; *p*-Methylstyrene-containing polyolefins

Crystallinity determination, variable temperature solid-state NMR, 167–168

Crystallization kinetics, in situ study of molecular, 173–175

Cyclohexylmethyldimethoxysilane (CHMMS)
effect on polymer molecular weight, stereoregularity, and chain-end composition, 53*t*
effect on relationship between propene pressure and catalyst stereoselectivity, 57*t*
external donor in magnesium chloride-based catalyst system, 53, 54

Cyclopentadienyltitanium tribenzyloxide (CpTi(OBz)₃). *See* Monocyclopentadienyltitanium/MAO catalysts

D

Dicyclopentyldimethoxysilane (DCPMS)
effect on polymer molecular weight, stereoregularity, and chain-end composition, 53*t*
effect on relationship between propene pressure and catalyst stereoselectivity, 57*t*
external donor in magnesium chloride-based catalyst system, 53, 54

1,3-Diene polymerization. *See* Lanthanide catalysts for 1,3-diene polymerization

Dipolar dephasing experiment, branch locations in polyolefins, 170–172

poly(propylene) by CpTi(OBz)₃/MAO1 system, 86
preparation of MAOs, 82
propylene and styrene homopolymerizations, 83–84
tail-to-tail and head-to-head enchainments, 88
tail-to-tail structure of Pr–St sequence by 2,1-insertion of St, 88

N

Neodymium catalysts
activity of compounds containing Cl atoms and Nd-allyl bonds, 19
catalyst activity and polymer *cis* content, 16
catalysts based on Nd-allyl compounds, 17, 19–20
conventional systems, 16–17
effect of aging time on activity, 18*f*
2,4-hexadiene polymers, 27
influence of B(C₆F₅)₃ on butadiene polymerization with allyl Nd catalysts, 22*f*
influence of order of catalyst component addition on heterogeneity and activity, 16–17
methods for preparation, 16
most interesting among lanthanides, 16
nature of catalytic species, 21
polymerization of 1,3-dienes, 15
polymerization of butadiene, 19–20
polymerization of butadiene with AlEt₂Cl–Nd(OCOR)₃–Al(iBu)₃ in different solvents, 22*f*
polymerization of substituted butadienes, 21, 23–27
polymer microstructure, 21, 23–25
polymer molecular weight from substituted butadienes, 25
polymers from terminally substituted monomers, 25, 27
polymers with terminally substituted monomers, 25, 27
scheme for formation of poly[(E)-1,3-pentadiene], 25, 26*f*
scheme for formation of poly[(E)-2-methyl-1,3-pentadiene], 25, 26*f*
X-ray powder spectra of polymers from terminally substituted monomers, 28*f*
See also Lanthanide catalysts for 1,3-diene polymerization

O

1-Octene, terpolymerization with ethylene and *p*-methylstyrene, 116–119
Oxidation, reaction on polyolefin containing *p*-methylstyrene, 120

P

Pentadiene
polymerization with neodymium catalyst, 23, 25
scheme for formation of poly[(E)-1,3-pentadiene] with mixed *cis*-1,4/1,2 structure, 26*f*
PhSi(OEt)₃, effect on relationship between propene pressure and catalyst stereoselectivity, 57*t*
cis-Poly(butadiene), identified catalysts for synthesis, 15–16
Poly(ethylene) (PE). *See* Ethylene; Linear low-density polyethylene (LLDPE)
Polyethylene product design
CCLDI quantifying distribution of chain segment lengths between branch points, 96
correlation between Relaxation Spectrum Index (RSI) and extruder motor load, 99*f*
ethylene polymer examples, 95–96
experimental, 95–96
EXXPOL mLLDPEs blown film performance, 98, 101
focus of technology developments, 101, 103
global demand for polyethylene (PE), 94
linear-low density polyethylene (LLDPE) types (I, II, III) using EXXPOL metallocenes, 94–95
molecular weight distributions by size exclusion chromatography (SEC), 95
polydispersity index-CCLDI property map for various PE samples, 96, 97*f*
processability characteristics of type III mLDPEs, 101
processability-toughness map cast in terms of RSI and dart impact, 100*f*
processability-toughness relations for various PE, 96*f*
RSI by small-amplitude dynamic oscillatory melt rheological tests, 95
RSI correlation with extrusion, ease/maximum output and bubble stability in high rate blown film, 98, 99*f*
RSI-processability correlation with melt tension, 99*f*
state-of-the-art methodologies, 95
temperature rising elution fraction (TREF) for inter-chain short chain branching distribution (SCBD) or CD, 95
TREF distributions for type I, II, III PE versus conventional Ziegler–Natta LLDPE, 96, 97*f*
type II film property evaluation, 101*t*
type III film property evaluation, 102*t*
type III product processability, 102*t*

RETURN TO ➤

CHEMISTRY LIBRARY
100 Hildebrand Hall • 642-3753

LOAN PERIOD 1	2	3
4	5 **1 MONTH**	6

ALL BOOKS MAY BE RECALLED AFTER 7 DAYS
Renewable by telephone

DUE AS STAMPED BELOW

FORM NO. DD5

UNIVERSITY OF CALIFORNIA, BERKELEY
BERKELEY, CA 94720-6000